Recent Advances in Biometrics

Edited by Muhammad Sarfraz

Published in London, United Kingdom

Recent Advances in Biometrics
http://dx.doi.org/10.5772/intechopen.97986
Edited by Muhammad Sarfraz

Contributors
Masashi Nishiyama, Hind Alrubaish, Nazar Saqib, Mridula Sharma, Haytham Elmiligi, Iannis Kominis, Michail Loulakis, Özgur E. Müstecaplıoğlu, Siddharth B. Baburao Dabhade, Nagsen S. Samadhan Bansod, Yogesh S. Rode, Narayan P. Bhosale, Prapti D. Deshmukh, Karvhari V. Kale, Eugene Fedorov, Tetyana Yuriyvna Utkina, Tetyana Neskorodieva, Taban Habibu, Anael Elikana Sam, Edith Talina Luhanga, David Palma, Pier Luca Montessoro, Mahmoud M.S. Mahrous Sayed Farrag, Huiqi Yvonne Lu, Muhammad Sarfraz, Nourah Alfialy

First published in London, United Kingdom, 2022 by IntechOpen
IntechOpen is the global imprint of INTECHOPEN LIMITED, registered in England and Wales, registration number: 11086078, 5 Princes Gate Court, London, SW7 2QJ, United Kingdom

British Library Cataloguing-in-Publication Data
A catalogue record for this book is available from the British Library

Additional hard and PDF copies can be obtained from orders@intechopen.com

Recent Advances in Biometrics
Edited by Muhammad Sarfraz
p. cm.
Print ISBN 978-1-80355-456-3
Online ISBN 978-1-80355-457-0
eBook (PDF) ISBN 978-1-80355-458-7

For EU product safety concerns:
IN TECH d.o.o., Prolaz Marije Krucifikse Kozulić 3, 51000 Rijeka, Croatia,
info@intechopen.com or visit our website at intechopen.com.

Meet the editor

Muhammad Sarfraz is a professor in the Department of Information Science, Kuwait University. His research interests include computer graphics, computer vision, image processing, machine learning, pattern recognition, soft computing, data science, intelligent systems, information technology, and information systems. Prof. Sarfraz has been a keynote/invited speaker on various platforms around the globe. He has advised various students for their MSc and Ph.D. theses. He has published more than 400 publications as books, journal articles, and conference papers. He is a member of various professional societies and a chair and member of the International Advisory Committees and Organizing Committees of various international conferences. Prof. Sarfraz is also an editor-in-chief and editor of various international journals.

Contents

Preface **IX**

Chapter 1 **1**
Introductory Chapter: On Biometrics with Iris
by Muhammad Sarfraz and Nourah Alfialy

Chapter 2 **23**
Biometric-Based Human Recognition Systems: An Overview
by David Palma and Pier Luca Montessoro

Chapter 3 **45**
Assessment of How Users Perceive the Usage of Biometric Technology Applications
by Taban Habibu, Edith Talina Luhanga and Anael Elikana Sam

Chapter 4 **69**
Behavioral Biometrics: Past, Present and Future
by Mridula Sharma and Haytham Elmiligi

Chapter 5 **89**
Biometrics of Aquatic Animals
by Mahmoud M.S. Farrag

Chapter 6 **115**
MedMetrics: Biometrics Passports in Medical and Clinical Healthcare That Enable AI and Blockchain
by Huiqi Yvonne Lu

Chapter 7 **131**
Quantum Biometrics
by Iannis Kominis, Michail Loulakis and Özgur E. Müstecaplıoğlu

Chapter 8 **149**
Feature Extraction Using Observer Gaze Distributions for Gender Recognition
by Masashi Nishiyama

Chapter 9 165
Image Acquisition for Biometric: Face Recognition
by Siddharth B. Dabhade, Nagsen S. Bansod, Yogesh S. Rode, Narayan P. Bhosale, Prapti D. Deshmukh and Karbhari V. Kale

Chapter 10 177
Your Vital Signs as Your Password?
by Hind Alrubaish and Nazar Saqib

Chapter 11 195
A Voice Signal Filtering Methods for Speaker Biometric Identification
by Eugene Fedorov, Tetyana Utkina and Tetyana Neskorodeva

Preface

A biometric system is a technological system that uses information about a person or other biological organism to identify that person. The biometric industry is rapidly changing and progressing. What used to be a futuristic concept has now become a reality. To work correctly and effectively, biometric systems depend and rely on data about specific biological traits. Biometric systems are widely used in various real-life applications. These include personal recognition, identification, verification, and more. Biometric systems may also be needed for safety, security, permission, banking, crime prevention, forensics, medical applications, communication, face finding, and so on.

The increasing trends, needs, and applications of biometric systems require new developments to achieve desired objectives. This involves capturing, storing, finding, retrieving, analyzing, and using biometrics in everyday life under the computing environment. Being a computer-based technology, biometric systems carry out automatic processing, manipulation, and interpretation of personal information. This book explores biometric systems education, research, applications, techniques, tools, and algorithms that originate from areas such as image processing, computer vision, pattern recognition, signal processing, artificial intelligence, intelligent systems, soft computing, computer engineering, electrical engineering, and computer science in general.

Chapters focus on the latest developments, theories, methods, approaches, algorithms, analysis, systems, hardware, and software for advancements in biometrics and related systems. It is a useful resource for professionals, researchers, academicians, engineers, scientists, and policymakers involved in biometrics.

Chapter 1, "Introductory Chapter: On Biometrics with Iris" by Sarfraz and Alfialy, provides an introduction to the topic with a focus on iris biometrics. Although there are various types of biometrics, such as fingerprint, face, speaker/voice, gait, keystroke, odor, and many more, much recent research has focused on the iris. As such, this chapter presents a comprehensive overview of iris recognition in biometric systems. It includes a comparative study as well as discusses future trends.

Chapter 2, "Biometric-Based Human Recognition Systems: An Overview" by Palma and Montessoro, provides an overview of the most used biometric traits along with their properties, the various biometric system operating modalities, and various related security aspects. It discusses the different stages of a biometric recognition process as well as the various threats that can compromise the security of a biometric system.

Chapter 3, "Assessment of How Users Perceivethe Usage of Biometric Technology Applications" by Habibu et al., discusses how biometrics transactions rely on end-user perceptions and responses. If end users are fearful, hesitant, or uneasy about biometric technology applications, misuse and implementation complications can occur. This chapter investigates end user motivation, understanding, consciousness, and acceptance of biometric technology applications via a public survey of 300 people in Uganda.

Chapter 4, "Behavioral Biometrics: Past, Present and Future" by Sharma and Elmiligi, discusses behavioral biometrics. Behavioral biometric authentication identifies users based on a set of unique behaviors that can be observed when users perform daily activities or interact with smart devices. This chapter discusses the different types of behavioral biometrics and explores various classifications of behavioral biometrics-based on their use models. The chapter highlights trending research directions in behavioral biometrics authentication and presents examples of current commercial solutions based on behavioral biometrics.

Chapter 5 "Biometrics of Aquatic Animals" by Farrag, discusses the use of biometrics in aquatic studies.

It presents research on biometrics of different aquatic animals, such as dolphins, sharks, rays, molluscs, crustaceans, protozoa, and so on, from different locations. Biometrics is considered an identifier for any new exotic or invasive species and it is the first step in biodiversity and stock management. This chapter also presents databases with some recent trends including animal biometric recognition systems for different applications and environments.

Chapter 6, "MedMetrics: Biometrics Passports in Medical and Clinical Healthcare That Enable AI and Blockchain" by Lu, introduces an emerging area of biometrics called MedMetrics, which combines medical and biological biometrics of patients based on their electronic health records, International Classification of Disease codes, time-series test results, and biological record to create coded "healthcare passports." This infrastructure can be used to identify patients, allowing healthcare providers and patients to access and add to encrypted patient records.

Chapter 7, "Quantum Biometrics" by Kominis et al. examines the human visual system's ability to perform efficient photon counting, which has been used to devise a new biometric authentication methodology. It presents and summarizes a recent proposal to use quantum light sources, particularly a single-photon source, to enhance the performance of the authentication process.

Chapter 8, "Feature Extraction Using Observer Gaze Distributions for Gender Recognition" by Nishiyama, studies the gaze distribution of observers viewing images of subjects for gender recognition. The authors propose a methodology that hypothesizes that the regions corresponding to the concentration of the observer gaze distributions contain discriminative features for gender recognition. Experimental results show that the observers mainly focused on the head region, not the entire body. Thus, gaze-guided feature extraction significantly improves the accuracy of gender recognition.

Chapter 9 "Image Acquisition for Biometric: Face Recognition" by Dabhade et al., discusses how to acquire face images using MATLAB. It considers image acquisition devices and image processes in the facial recognition process.

Chapter 10, “Your Vital Signs as Your Password?” by Alrubaish and Saqib, investigates the ability to use vital signs obtained via electrocardiogram (ECG) and electroencephalogram (EGG) as unimodal authentication. It highlights recent techniques, their requirements and limitations, and whether they are ready to be used in the real market. It discusses the applicability of unique study and observes that the vital signs can be considered as personal a PASSWORDpasswords due to their uniqueness and resistance to spoofing and other attacks. its uniqueness, but it needs more improvements to be deployed to the market.

Finally, Chapter 11, “Voice Signal Filtering Methods for Speaker Biometric Identification” by Eugene Fedorov et al., examines various methods of suppressing noise in a voice signal to be used for biometric identification. The chapter discusses several types of filtering methods and presents the results of numerical research of denoising methods for voice signals from the TIMIT database with additive Gaussian noise and multiplicative Gaussian noise.

Muhammad Sarfraz
Department of Information Science,
College of Life Sciences,
Kuwait University,
Sabah AlSalem University City, Shadadiya, Kuwait

Chapter 1

Introductory Chapter: On Biometrics with Iris

Muhammad Sarfraz and Nourah Alfialy

1. Introduction

Biometrics is the systematic study of measuring and analyzing biological data for the purpose of validation or identification. Biometrics refers to specific physiological and/or behavioral (extrinsic and/or intrinsic respectively) characteristics that are uniquely related to a person [1, 2]. The biometric systems use unique human physiological and anatomical properties to define details. Such systems effectively help to overcome the security issues affecting the conventional methods of personal authentication. In the smart world today, the importance of technological solutions in biometrics is growing. Specifically, automation culture has desired to design and launch automated systems for highly reliable and accurate human authentication and identification.

Biometrics has been deployed successfully in various fields of real life. Numerous methods, techniques, and systems on biometrics serve sciences, security, military, medical area, and human identification. There are various kinds of biometrics being used, these include Fingerprint, Face, Speaker/Voice, Infrared thermogram (facial, hand, or hand vein), Gait, Keystroke, Odor, Ear, Hand geometry, Retina, Iris, Palmprint, Signature, DNA, Knuckle crease, Brain/EEG, Heart sound/ECG. Specifically, in the past decade, iris recognition technology has become the most popular biometric technology for human authentication and recognition due to its stability and uniqueness in its structure. The iris has a unique structure that remains stable throughout a person's life. Iris recognition is one of the authentication methods that uses high-resolution assisted pattern recognition technology. The general method of the iris recognition system includes image acquisition, segmentation, feature extraction, matching, and classification [3].

Iris has become a very effective recording of its superior properties, such as reliability and accuracy. In recent years, a good amount of research is made regarding the evolution of biometric-based on the iris. This presented iris recognition as a very clear and effective concept. There is a need to highlight and analyze the work done by different authors related to iris studies, methodologies, and practices. A detailed comparative study could particularly provide an overview for the readers.

The idea of iris recognition goes back to an eighteenth-century Paris prison, where police distinguished criminals by examining the color of their irises. Daugman [4] was the first to develop the basic algorithms that now form the basis of all current commercial iris recognition systems, having been commissioned by Flom and Safir [5, 6] to conduct extensive and comprehensive research to implement automatic iris recognition. In 1987, Flom and Safir acquired a non-applied concept of an automated iris biometric system. A report was published by Johnston in 1992 without any experimental results [5, 6].

The motivation behind this work is to study biometric iris recognition specifically because it provides one of the most stable biometric signals to recognize distinct tissues that form prematurely and remain constant throughout life unless there is an eye injury.

The aim of this chapter is to contribute in a comprehensive survey about the difference between the existing biometrics techniques. An important aspect of biometric technology is to evaluate its performance. The performance of any biometric authentication technology can be measured by various parameters. Compared to other vital features, such as the face, fingerprint, and voice, the iris patterns are more stable and reliable [7–9]. The reason behind this is that iris recognition algorithms require pre-processing of the input image to obtain better data quality by tracking different feature points of the iris. Biometrics using a feature is so unique that the chance of any two people having the same features is very rare [10]. Identification of a person based on recognition of the iris of the eye gives one of the most reliable results. Iris tissue features provide unique high-dimensional information that explains why iris recognition-based verification has the lowest false acceptance rate among all types of identity verification systems [1, 11–13].

Iris recognition has been used in many countries with the purpose of identifying millions of people around the world. This technology is comfortable to use and difficult to rig. Many authentication programs, including border crossings without a passport, national identity, etc., have adopted this technology for its benefits [14]. For the purpose of human recognition, the iris biometric recognition system has proven its importance. The biometric iris recognition systems are easy to use and create a hassle-free security environment. Iris scanners can be used to protect high-value websites by blocking the access of unwanted visitors. Commercial and governmental institutions in all fields have recognized the benefits of this system and have embarked on implementing validation systems based on iris recognition in a major way [15]. Iris recognition is one of the best-protected methods of authentication and recognition. Iris recognition accuracy is very promising. The false acceptance rate, as well as the rejection rate, is very low. A special grayscale camera is used to take an iris pattern within 10–40 cm from the camera [14, 16–18]. An appropriate methodology is used to define the irises of the image, and if it is present, a grid of curves covering the iris is created and the iris symbol is generated based on the opacity of the points. It is affected by two things—first, the general opacity of the image, and secondly, the changes in the size of the iris. The comparison of two irises can be computed through the knock distance based on the difference in the number of bits and it is very fast [4, 19]. Also, the template matching technique can be used, and it uses statistical calculation to match the stored iris template and the obtained iris template. Iris recognition is applied in the following areas: border control, passports, ID cards, and other government purposes, database access, login authentication, aviation security, hospital security, access control to buildings, areas, homes, and security of restricted prisons [6, 20, 21].

For convenience, it is desired to know the basic concepts and terminologies we are going to use throughout this chapter. There are as follows:

- **Biometric:** Originated from the Greek words bios (life) and metric or (measure), directly translates into "life measurement" [22].

- **Iris:** It is a circular shape structure in the eye.

- **Iris normalization:** "Performed to convert the iris coordinates into polar coordinates to rectangular iris template to make it constant and persistent against the effect of changing the size of the pupil. Once the outer and inner circles of the iris are localized, these values are taken as input to the Daugman's Rubber-sheet model" [23].

- **Feature extraction:** "After pre-processing of the image, feature extraction is carried out on normalized iris image" [23].
- **Daugman's approach:** Daugman's patent states that "the system acquires through a video camera a digitized image of an eye of the human to be identified" [6].
- The iris boundary is explained with parameters, which are the radius and the coordinates of the center of the circle of the iris boundary. Daugman proposed the integrodifferential operator to detect the iris boundary by finding the parameter space [6].

This chapter has been organized in various sections. Section 1 gives a brief introduction about biometric iris recognition, motivation to work in this study, the importance of this study, basic concepts and terminologies to be used, and the organization of this study. Section 2 consists of a literature survey and a comparative study of the existing different methods used in biometric iris recognition. It also gives information about different methods used in the extraction of the features of iris image datasets and data analysis. Finally, Section 3 gives the new directions for the future. It suggests some recommendations for community, government, industry, etc. Then the overall conclusion of the study is done in this chapter. It concludes with the discussion on future trends as well.

2. Literature survey

Although many papers have been published in this field in the past years, twenty-six papers have been selected and presented to understand the iris recognition techniques available in the literature. These articles have shown differences between each other in one way or another. In this chapter, a review is presented focusing on all four phases, i.e., segmentation, normalization, extraction, and template algorithms of the iris recognition technology from Daugman's initial work in 1993 to some recent work.

Daugman [10] developed a feature extraction process based on information from a group of 2D Gabor filter. He created a file 256 bytes by specifying the local phase angle according to outputs of the real and imaginary parts of the filtered image, compare the percentage of mismatched bits between a pair of Iris representations via the XOR operator, and the choice of a separation point in the space of the Hamming distance.

On the contrary, Wildes system took advantage of the Laplacian pyramid, which was built with four different precision levels. Generate the iris symbol [14, 15]. Also, it explained a normalized correlation based on goodness-of-match values and Fisher's linear discriminant for pattern matching. Both iris recognition systems use of bandpass image decompositions to get multi-scale information.

Lim et al. [21] proposed an iris recognition system. It includes a compact representation scheme for iris patterns by the 2D wavelet transform. This method is used for initializing weight vectors and determining winners for recognition in a competitive learning method. Flom and Safir [6] had earned a patent in Iris

Recognition System, which gives a generalized concept in using iris as a biometric system but does not describe any implemented algorithm.

In the process of recognizing the iris of the eye, conversion is necessary. An iris image acquired in a convenient symbol can easily manipulate it. Hence, we will take a quick look at the process of feature extraction and representation of modern wonderful works and papers. Iris recognition is the procedure of comparing known and unknown irises to prove that it is from the same person or not [11].

Today, many approaches, techniques, and systems are used to match iris and solve related problems. This section is focused on analyzing and categorizing different author's work in the iris recognition area. **Table 1** provides a summary of various papers in the current literature. First column determines the Reference of the papers by author names and year of publication. Second column gives the summary of the work in the corresponding paper, and the third column describes the implemented approaches used to solve iris recognition issues. The author names and the year of publication have been used as an identifier for the rest of the tables in the chapter showing other details of the referred literature.

The main point of biometrics technology is to evaluate their performance and accuracy. It can be measured by the various parameters such as False Accept Rate (FAR), False Reject Rate (FRR) and Crossover Rate (CER) or Equal Error Rate (EER). An identity claims wrongly rejected is called False Rejection and a false identity claims wrongly accepted is known as False Acceptance. In order to make limited entry to authorized users FAR and FRR are used. False Rejection Rate (FRR) measures the probability of rejecting an authorized user incorrectly as an invalid user [16].

Table 2 shows the accuracy and performance in percentage. It also mentions the identification and verification measures. Identification and verification are matching techniques for Iris recognition. In the verification, the person enrolls his Iris to the system and the template is stored in the database. Every time the person accesses the system, he has entered his iris to verify himself. It's a one-to-one relationship where the input Iris is compared with the stored one. On the other hand, identification is one to many relationships because the human Iris is matched with the Irises in 7the database to determine who is that person [11]. While the performance measures used for identification depend on the accuracy, recognition rate, rank K, etc., the performance measures for verification are False Match Rate (FMR), False Non-Match Rate (FNMR), False Accept Rate (FAR), False Rejection Rate (FRR), and Equal Error Rate (EER). The researchers in [3] describe the meaning of the authentication parameters. FAR happens when the system recognizes person erroneous. But when the system rejects entry to approve person that means the FRR is happening. FMR is the amount of fraud assessments with threshold value "T" divided by the total quantity of fraud similarities. FNMR is the quantity with unaffected comparisons with threshold value "T" divided by the total quantity of open comparisons. Last one is EER; it describes the error rate of the system.

In Lim et al. [21], eye images captured at a distance with the help of a CCD camera. Then, in the acquired image, iris is segmented. Initially, it is done by detecting the pupil using the center point detection method followed by edge detection method by finding virtual circles. An analysis was made in the pre-processing stage, with 6000 data to identify the causes of failure at this stage. Data involved images both with Lens, without, and with glasses. In the normalization stage a 450×60 bit iris image part was obtained. Gabor transforms and Haar wavelet transforms, which are two different methods, were used to analyze and extract the features from the segmented iris image.

Reference	Brief summary of the paper	Approaches adopted
Choudhary, Tiwari and Singh, 2012 [3]	Available feature extraction methods for iris pattern are studied in this paper. This paper is an analysis of the result of the various feature extraction methods. Iris localization using Hough transform performs better as compared to other localization techniques in case of occlusion due to eyelids and eyelashes.	• Corner Detection Based Iris Encoding. • Feature extraction using Haar wavelet. • Feature extraction using Gabor filter. • Statistical pattern recognition • Multichannel Gabor Filter
Rakesh and Khogare, 2012 [5]	In the feature extraction process, Gabor wavelet and wavelet transform, which are widely used for extracting features, were evaluated. From this evaluation, they found that Haar wavelet transform has better performance than that of Gabor transform. Second, Haar wavelet transform was used for optimizing the dimension of feature vectors in order to reduce processing time and space. They could present an iris pattern without any negative influence on the system performance. Last, they improved the accuracy of a classifier, a competitive learning neural network, by proposing an initialization method of the weight vectors and a new winner selection method designed for iris recognition. With these methods, the iris recognition performance increases to 98.4%.	• Independent Component Analysis. • Multichannel Gabor filtering and 2D wavelet transforms. • Zero-crossing Representation Method • Iris Recognition Using Cumulative-Sum-Based Change Analysis. • Iris Recognition through Improvement of Feature vector and classifier
Arrawatia, Mitra and Kishore, 2017 [23]	This paper offers review on existing technologies for iris recognition proposed by various researchers. Iris localization and segmentation, wavelets are used impressively, and Gabor filters are used for coding. There are two other popular techniques for segmentation: canny edge detector and Hough transform. But after adding Contourlet transform with the Hough transform and canny edge detector gives better results in segmentation which rates up to 100 percent. The comparison of result shows that this method for segmentation gives much better result for iris image segmentation with high accuracy and efficiency, which maintain the basic quality of image. For iris normalization, Daugman's rubber-sheet model achieves better result by reducing dimensional inconsistencies.	• Image Capturing/ Acquisition • Iris Segmentation and Localization • Iris Image Denoising by Contourlet Transform • Normalization Stage • Feature Extraction • Feature Coding • Matching Algorithm
Bowyer, Hollingsworth and Flynn, 2008 [6]	This survey suggests a structure for the iris biometrics literature and summarizes the current state-of-the art. Most research publications can be categorized as making their primary contribution to one of the four major modules in iris biometrics: image acquisition, iris segmentation, texture analysis and matching of texture	Flom and Safir's concept patent • Daugman's approach • Wildes' approach

Reference	Brief summary of the paper	Approaches adopted
	representations. Other important research includes experimental evaluations, image databases, applications and systems, and medical conditions that may affect the iris.	
Sheela and Vijaya, 2010 [18]	In this paper, different iris recognition methods, which aid an appropriate outlook for future work to build integrated classifier on the latest input devices for excellent business transactions, are discussed. Benchmark databases, products are also discussed. Since the area is currently one of the most on the go and the bulk of research is very large, this survey covers some of the significant methods.	• Phase-based method • Texture analysis based method • Zero-crossing representation method • Approach based on intensity variations
Sanjay, Ganorkar, Ashok and Ghatol, 2007, 2004 [20]	This paper presents a literature survey on the various techniques involved in identification and the emphasis given on biometric recognition system. In various applications, the biometric recognition system has been proved to be accurate and very effectively.	• Sensor Module/Image Acquisition • Feature Extraction Module • Database Module • Matching Module
Daugman and Downing, 2001 [10]	This paper investigated the randomness and uniqueness of human iris patterns by mathematically comparing 2.3 million different pairs of eye images. The phase structure of each iris pattern was extracted by demodulation with quadrature wavelets spanning several scales of analysis. The resulting distribution of phase sequence variation among different eyes was precisely binomial, revealing 244 independent degrees of freedom.	• Complex-valued two-dimensional (2D) • Gabor Wavelets • The phase-quadrant demodulation process
Daugman, 2004 [11]	Algorithms developed by the author for recognizing persons by their iris patterns have now been tested in many field and laboratory trials, producing no false matches in several million comparison tests. The recognition principle is the failure of a test of statistical independence on iris phase structure encoded by multi-scale quadrature wavelets. The combinatorial complexity of this phase information across different persons spans about 249 degrees of freedom and generates a discrimination entropy of about 3.2 b mm2 over the iris, enabling real-time decisions about personal identity with extremely high confidence.	• Demodulation • Focus Assessment • Gabor wavelets
Sanjay, Ganorkar, Ashok and Ghatol, 2007 [20]	In this paper, the system steps are capturing iris patterns; determining the location of iris boundaries; converting the iris boundary to the stretched polar coordinate system; extracting iris code based on texture analysis. The system has been implemented and tested using dataset of number of samples of iris data with different contrast quality. The developed algorithm performs	• Binary Segmentation • Pupil Center Localization • Circular Edge Detection • Remapping of the Iris

Reference	Brief summary of the paper	Approaches adopted
	satisfactorily on the images, provides 93% accuracy. Experimental results show that the proposed method has an encouraging performance.	
Anil, Ross and Prabhakar, 2004 [22]	In this paper, a brief overview of the field of biometrics is given, and it summarizes some of its advantages, disadvantages, strengths, limitations, and related privacy concerns.	• Sensor Module • Feature Extraction Module • Matcher Module • System Database Module
Bramhananda, Reddy and Goutham, 2018 [12]	This paper throws light into various iris-based biometric systems, issues with iris in the context of texture comparison, cancellable biometrics, iris in multimodel biometric systems, iris localization issues, challenging scenarios pertaining to accurate iris recognition and so on.	• Hamming Distance Classifier (HDC) for predicting False Rejection Rate (FRR) and False Acceptance Rate (FAR)
Roy and Bandyopadhyay, 2017 [1]	This paper highlighted the detection of iris using biotechnology technique.	—
Phadke, 2013 [13]	This paper discussed various Biometric Identification Systems which can be grouped based on the main physical characteristic that lends itself to biometric identification; Fingerprint identification, Hand geometry, Palm Vein Authentication, Retina scan, Iris scan, Face recognition, Signature, Voice analysis.	—
Rui and Yan, 2018 [24]	In this paper, the authors classified and thoroughly review the existing biometric authentication systems by focusing on the security and privacy solutions. They had analyzed the threats of biometric authentication and proposed several criteria with regard to secure and privacy-preserving authentication. They had further reviewed the existing works of biometric authentication by analyzing their differences and summarizing the advantages and disadvantages of each based on the proposed criteria. This paper discussed the problems of aliveness detection and privacy protection in biometric authentication.	—
Jin-Hyuk, Eun-Kyung and Sung-Bae, 2004 [7]	This paper gives a comprehensive overview of biometric technology and performance evaluation with more than 100 publications. After the thorough review, it proposed a promising evaluation method based on affecting factors.	—
Manisha and Kumar, 2019 [8]	This research paper presented a comprehensive survey of more than 120 techniques suggested by various researchers for Cancelable Biometrics and a novel taxonomy for the same is developed. Further, various performance measures used	• Cryptography based methods • Transformation based methods • Filter based methods • Hybrid methods

Reference	Brief summary of the paper	Approaches adopted
	in Cancelable Biometrics are reviewed and their mathematical formulations are given. It also suffers from various security attacks as given in literature. A review of these security attacks is carried out. It also performed a review of databases used in literature for nine different Cancelable Biometrics.	• Multimodal based methods
Himanshu, 2013 [9]	This paper proposed a personal identification using iris recognition system with the help of six major steps, which are image acquisition, localization, isolation, normalization, feature extraction, and matching, and these six steps consists several minor steps to complete each step. The boundaries of the iris, as papillary and limbic boundary, are detected by using Canny Edge Detector & Circular Hough Transformation. It used masking technique to isolate the iris image from the given eye image, this isolated iris image is transformed from Cartesian to polar coordinate. Finally extract the unique features of the iris after enhancing the iris image and then perform matching process on iris code using Hamming Distance for acceptance and reject process.	• Canny Edge Detector & Circular Hough Transformation • Localization • Enhancement and Denoising • Feature Extraction
Williams, 1997 [25]	This paper discussed Iridian Technologies systems that have enrolled 99.99% of the irises presented to them. This Technologies iris recognition system has allowed no False Accept errors in over three million file comparisons during the two testing programs referenced in this paper, and in millions of file comparisons elsewhere. Under the formal, controlled DOD testing scenario, the system was 99.95% accurate in the area of False Rejects, with only one False Reject out of 1,995 trials. The reason for that error was identified, corrected, and never repeated.	• Hamming Distance Calculation • Recognition or Rejection
Wildes, 1997 [15]	This paper examines automated iris recognition as a biometrically based technology for personal identification and verification. The motivation for this endeavor stems from the observation that the human iris provides interesting structure on which to base a technology for noninvasive biometric assessment.	• Image Acquisition • Iris Localization • Pattern Matching
Wildes, Asmuth, Green, Hsu, Kolczynski, Matey and McBride, 1994 [14]	This paper describes a prototype system for personnel verification based on automated iris recognition. The motivation for this endeavor stems from the observation that the human iris provides a particularly interesting structure on which to base a technology for noninvasive biometric measurement.	• Image Acquisition • Iris Localization • Pattern Matching

Reference	Brief summary of the paper	Approaches adopted
Ganorkar and Ghatol, 2007 [19]	In this paper, iris recognition as one of the important method of biometrics-based identification systems and iris recognition algorithm is described. The system steps are capturing iris patterns; determining the location of iris boundaries; converting the iris boundary to the stretched polar coordinate system; extracting iris code based on texture analysis. The system has been implemented and tested using dataset of number of samples of iris data with different contrast quality.	• Binary Segmentation • Pupil Center Localization • Circular Edge Detection • Remapping of the Iris
Daugman, 1993 [4]	This paper studies the method for rapid visual recognition of personal identity based on the failure of a statistical test of independence. The most unique phenotypic feature visible in a person's face is the detailed texture of each eye's iris. The visible texture of a person's iris in a real-time video image is encoded into a compact sequence of multi-scale quadrature 2-D Gabor wavelet coefficients, whose most-significant bits comprise a 256-byte. Statistical decision theory generates identification decisions from Exclusive-OR comparisons of complete iris codes at the rate of 4000 per second, including calculation of decision confidence levels.	• Complex-valued two-dimensional (2D) • Gabor wavelets • The phase-quadrant demodulation process
Lim, Lee, Byeon and Kim, 2001 [21]	This paper proposed an efficient method for personal identification by analyzing iris patterns that have a high level of stability and distinctiveness. In order to improve the efficiency and accuracy of the proposed system, it presented a new approach to making a feature vector compact and efficient by using wavelet transform, and two straightforward but efficient mechanisms for a competitive learning method such as a weight vector initialization and the winner selection. With all these novel mechanisms, the experimental results showed that the proposed system could be used for personal identification in an efficient and effective manner.	• Image Acquisition • Pre-processing Stage • Feature Extraction Stage • Identification and Verification Stage • Wavelet Transform • Gabor Transform

Table 1.
An overview of literature.

In Daugman [11], proposed an approach that is an improvement to his previous work. This approach is working with the noise disturbances that occur while acquiring an iris image of a human eye. Also, an algorithm was introduced for detecting the eyelids, which involves arcuate edges with spline parameter, instead of circular edges in the Integro differential operator.

In Wilde [14, 15] tried a different approach, in which inner and outer iris boundary is computed with the help of a gradient-based binary edge map followed by circular

Reference	Accuracy (Performance)	Performance measures used for verification	Performance measures used for identification
Choudhary, Tiwari and Singh, 2012 [3]	• 95% (Singh et al.) • 95.4% (Gupta et al.) FAR/FRR: 4/5 • 96.3% (Greco et al.) FAR/FRR: 3/4 • 94.85% (Tuama) 2.43/3.17 • 95.68% (Li Ma)	FAR, FRR	Accuracy, Recognition Rate, Iris Normalization, Pattern Matching, Feature Extraction, and Training & Testing Time.
Rakesh and Khogare, 2012 [5]	• 98.4%	EER	Biometric, Gabor filtering, wavelet transform, cumulative sum, zero-crossing and feature vector.
Arrawatia, Mitra and Kishore, 2017 [23]	• 100%	FAR, FRR	Image Acquisition, Localization, Segmentation, Normalization, Feature Extraction, Template Generation, Pattern Matching.
Bowyer, Hollingsworth and Flynn, 2008 [6]	• Daugman's algorithm performed the best with 99.90% accuracy • Ma's algorithm with 98.00% • Avila's algorithm with 97.89% • Tisse's algorithm with 89.37%	FAR,FRR, EER	True Accept, False Accept, Receiver Operating Characteristic, Cumulative Match Characteristic
Sheela and Vijaya, 2010 [18]	• The experiments were conducted on UBIRIS database with accuracy of 98.02 and 97.88% for images captured in session 1 and session 2, respectively. • The segmentation performance for 1214 good quality images and 663 noisy images was 98.02 and 97.88%, respectively.	FAR, FRR	Phase-based method, Texture-analysis, Zero-crossing, Local intensity variations, Independent Component Analysis, Continuous Dynamic Programming.
Sanjay, Ganorkar, Ashok and Ghatol, 2007, 2004 [20]	Iris Recognition Performance Evaluation: • 0.94% (FAR) • 0.99% (FRR) • 0.01% (CER) • 0.50% (FTE)	FAR, FRR, CER, EER	False Rejection, False Acceptance, Face Recognition Vendor Test (FRVT).
Daugman and Downing, 2001 [10]	• 95%	—	—
Daugman, 2004 [11]	—	—	—
Sanjay, Ganorkar, Ashok and Ghatol, 2007 [20]	• 93%	—	—
Anil, Ross and Prabhakar, 2004 [22]	• 99%	FER, FTE, FTC, FNMR, FMR	False Non-Match Rate, False Match Rate

Reference	Accuracy (Performance)	Performance measures used for verification	Performance measures used for identification
Bramhananda, Reddy and Goutham, 2018 [12]	—	—	—
Roy and Bandyopadhyay, 2017 [1]	—	—	—
Phadke, 2013 [13]	—	—	—
Rui and Yan, 2018 [24]	—	FAR, FRR, EER	—
Jin-Hyuk, Eun-Kyung and Sung-Bae, 2004 [7]	—	FAR, FRR, FMR	FTE, FTA, FMR/FNMR, FAR/FRR, and FR for each trial are used to estimate the recognition performance, and processing time, efficiency of matching algorithm, performance for a specific population are used to analyze the results.
Manisha and Kumar, 2019 [8]	—	FAR, FPR	Failure to Acquire Rate (FTAR), Failure to Capture Rate (FTCR), FMR and FTA
Himanshu, 2013 [9]	• 98.9%	—	—
Williams, 1997 [25]	• 99.95%	—	Crossover (Equal) Error Rate (CER), Recognition Speed, Enrollment, Confidence, Testing
Wildes, 1997 [15]	—	—	—
Wildes, Asmuth, Green, Hsu, Kolczynski, Matey and McBride, 1994 [14]	—	—	—
Ganorkar and Ghatol, 2007 [19]	• 93%	—	—
Daugman, 1993 [4]	—	—	—
Lim, Lee, Byeon and Kim, 2001 [21]	• 97.1% (Learning Data) • 95.9% (Test Data) • Overall Performance 97.1–98.4%.	FAR, FPR	Learning Data, Test Data, Wavelet transform, Gabor transform, Multi-dimensional Winner Selection, Euclidean distance-based winner selection

Table 2.
An overview of literature for Accuracy and performance.

Hough transform. Wilde used around 60 human irises captured from 40 subjects in his experiment. Also, he has done a comparative study with Daugman's work in his paper. This method proved to provide higher accuracy rate when tested in CASIA database.

These algorithms can provide rotation, translation and size invariant result. Simulation results of these algorithm prove to provide a higher correct accept and reject rate. Results were tested using CASIA database, UBIRIS, UPOL, MMU and a database provided by Institute of Automation for 2005 Biometrics Authentication competition.

Reference	**Biometric**	**Database**	**No. of identities**
Choudhary, Tiwari and Singh, 2012 [3]	• Iris	• CASIA	• 1024
Rakesh and Khogare, 2012 [5]	• Iris	• UBIRIS • CASIA-IrisV3 • ND 2004–2005 database	• 241 • 1500
Arrawatia, Mitra and Kishore, 2017 [23]	• Iris	• CASIA • LEI • UPOL	—
Bowyer, Hollingsworth and Flynn, 2008 [6]	• Iris	• CASIA 1 • CASIA 3 • ICE2005 • ICE2006 • UBIRIS • UPOL	• 108 • 1500 • 244 • 480 • 241 • 128
Sheela and Vijaya, 2010 [18]	• Iris	• UBIRIS V1 • UBIRIS V2 • CASIA V1 • CASIA V2 • CASIA V3-Interval • CASIA V3-Lamp • CASIA V3-Twins • ND 2004–2005 • Iris DB 100 • Iris DB 800 • Iris DB 1600 • UPOL • MMU1 • MMU2	• 241 • 261 • 108 • 60 • 249 • 411 • 200 • 356 • 200 • 400 • 800 • 61 • 100 • 100
Sanjay, Ganorkar, Ashok and Ghatol, 2007, 2004 [20]	• Fingerprint • Face • Retina • Iris • Hand Geometry • DNA • Ear • Body Odor • Palm Print • Lip Motion • Hand Vein • Gait • Signature • Voice	—	—

Reference	Biometric	Database	No. of identities
Daugman and Downing, 2001 [10]	• Iris	—	—
Daugman, 2004 [11]	• Iris	—	—
Sanjay, Ganorkar, Ashok and Ghatol, 2007 [20]	• Iris	• CASIA	—
Anil, Ross and Prabhakar, 2004 [22]	• Iris • DNA • Face • Hand and finger geometry • Fingerprint • Signature • Voice • Retinal scan	—	—
Bramhananda, Reddy and Goutham, 2018 [12]	• Iris • Face • Fingerprint	—	—
Roy and Bandyopadhyay, 2017 [1]	• Iris	—	—
Phadke, 2013 [13]	• DNA • Fingerprint • Hand Geometry • Hand Vein • Iris Face/Facial Thermo Gram • Physiological • Retinal Scan • Signature • Voice	—	—
Rui and Yan, 2018 [24]	• Iris • Fingerprint • Voice • Keystroke • Face	—	—
Jin-Hyuk, Eun-Kyung and Sung-Bae, 2004 [7]	• Iris • Fingerprint	—	—
Manisha and Kumar, 2019 [8]	• Fingerprint	—	—
Himanshu, 2013 [9]	• Iris	• CASIA • MMU	—
Williams, 1997 [25]	• Iris	—	—
Wildes, 1997 [15]	• Iris	—	—
Wildes, Asmuth, Green, Hsu, Kolczynski, Matey and McBride, 1994 [14]	• Iris	—	—
Ganorkar and Ghatol, 2007 [19]	• Iris	• CASIA	—
Daugman, 1993 [4]			
Lim, Lee, Byeon and Kim, 2001 [21]			

Table 3.
An overview of literature for data used.

The experimental parts of the author's [1–25] are shown in **Table 3**. It explains the type of applications and kind of Databases used. Then, it shows the number of data used in the study.

There is a need to overlook for the data images together with their resolution and format. **Table 4** describes the number of data used in the application, the number of images resulted, their resolutions and formats.

In [10], Daugman described how iris recognition is being used to check visitors coming to the United Arab Emirates (UAE) against a watch-list of people who are denied entry to this country. The UAE database contains around 632,500 different iris images. In all comparison, no false matches were found with Hamming distances below about 0.26. Daugman reports that "to date, some 47,000 persons have been caught trying to enter the UAE under false travel documents, by this iris recognition

Reference	Total no. of images	Resolution (in pixels)	Image format
Choudhary, Tiwari and Singh, 2012 [3]	• 60	• 80×360	—
Rakesh and Khogare, 2012 [5]	• 1877 • 22,051 • 6000	• 64×300	—
Arrawatia, Mitra and Kishore, 2017 [23]	—	• 10 pixels and angular resolutions with angles varying from 00 to 3600	—
Bowyer, Hollingsworth and Flynn, 2008 [6]	• 756 • 22,051 • 2953 • 60,000 • 1877 • 384	—	—
Sheela and Vijaya, 2010 [18]	• 1877 • 11,102 • 756 • 1200 • 2655 • 16213 • 3183 • 64,980 • 8,000 • 16,000 • 32,000 • 384 • 450 • 995	• 400×300 • 800×600 • 320×280 • 640×480 • 320×280 • 640×480 • 640×480 • 640×480 • 1280×960 • 1280×960 • 1280×960 • 576×768 • 320×280 • 320×280	• jpeg • jpeg • bmp • bmp • jpeg • jpeg • jpeg • tiff • bmp • bmp • bmp • png • bmp • bmp
Sanjay, Ganorkar, Ashok and Ghatol, 2007, 2004 [20]	—	—	—
Daugman and Downing, 2001 [10]	• 2150	• 640×480	—
Daugman, 2004 [11]	—	• 640×480	—
Sanjay, Ganorkar, Ashok and Ghatol, 2007 [20]	• 51	—	—
Anil, Ross and Prabhakar, 2004 [22]	—	—	—
Bramhananda, Reddy and Goutham, 2018 [12]	—	—	—

Reference	Total no. of images	Resolution (in pixels)	Image format
Roy and Bandyopadhyay, 2017 [1]	—	—	—
Phadke, 2013 [13]	—	—	—
Rui and Yan, 2018 [24]	—	—	—
Jin-Hyuk, Eun-Kyung and Sung-Bae, 2004 [7]	—	—	—
Manisha and Kumar, 2019 [8]	—	—	—
Himanshu, 2013 [9]	—	—	—
Williams, 1997 [25]	—	—	—
Wildes, 1997 [15]	—	—	—
Wildes, Asmuth, Green, Hsu, Kolczynski, Matey and McBride, 1994 [14]	—	—	—
Ganorkar and Ghatol, 2007 [19]	• 51	—	—
Daugman, 1993 [4]			
Lim, Lee, Byeon and Kim, 2001 [21]	• 6000	• 450×60	

Table 4.
An overview of literature for data images with resolution and format.

Reference	Application	Reason of application	Advantages	Disadvantages
Choudhary, Tiwari and Singh, 2012 [3]	—	—	—	—
Rakesh and Khogare, 2012 [5]	• Computer System Security • Secure Electronic Banking • Mobile phones • Credit cards	• Iris patterns have stable, invariant, and distinctive features for personal identification.	—	—
Arrawatia, Mitra and Kishore, 2017 [23]	• Finance and banking • Healthcare and welfare • Immigration and border control • Public safety • Point of sale and ATM • Hospitality and tourism	• For verification and identification • Identify the accurate patient. • For security purpose the iris recognition technique is used in many countries borders and airports. • Some law enforcement agencies save the	• It is less time consuming and improves the standard of service and the customer or user will free from document verification process for identification, which is more time consuming. • Provides a high accuracy then	—

Reference	Application	Reason of application	Advantages	Disadvantages
		criminal data to track them. • The vulnerable Pos terminal is hacked by a hacker for the regular payment. For this activity, they are using the skimmers. These skimmers are installed at terminals which read and transmit the information of swiped card. • To overcome the unwanted access of a user in hotel room.	other biometric technique and removal of delicacy in medical record of a person. • Utilizing the security and accuracy of iris recognition system these agencies track the terrorist & criminals. • Law enforcement agencies use the saved biometric data of criminal record to enhance the security of public. • Iris recognition system used on all swipe or ATM machines so that hacker never use the information of others.	
Bowyer, Hollingsworth and Flynn, 2008 [6]	• "EyeCert" system	• Issue identity cards to authorized users. The barcode on the cards would store both biometric information about the person's iris, as well as other information, such as a name, expiration date, birth date, and so forth.	• The system is designed to allow identity verification to be done offline, thus avoiding potential problems that would come with systems that require constant access to a centralized database.	—
Sheela and Vijaya, 2010 [18]	• Civilian Identification management program • The Offender Identification System [Offender-ID] • PIER 2.4 • The Handheld Interagency Identity Detection Equipment [HIIDE]	• Iris authentication product • Supports identification of prisoners in jail environment. • Provides mobile identification with iris technology in a real-time environment. • Multi-biometric handheld device. It is used in defense agencies and in	—	—

Reference	Application	Reason of application	Advantages	Disadvantages
	• The LG Iris Access, Panasonic BM-ET200, Oki, IBM, Iris Guard IG-AD100, Sage, Spectrometric and Argus systems	remote or centralized enrolments • Work by analyzing the iris patterns and converting them into digital templates.		
Sanjay, Ganorkar, Ashok and Ghatol, 2007, 2004 [20]	• Border Control • Passports and Identity Cards. • Database Access • Login Authentication • Aviation Security • Hospital Security • Controlling access to restricted buildings, areas, homes and prison security	• Iris recognition is one of the best-protected approaches for authentication and recognition,	• The accuracy of Iris recognition is most promising. • The false acceptance rate as well as rejection rate is very low.	—
Daugman and Downing, 2001 [10]	—	—	—	—
Daugman, 2004 [11]	—	—	—	—
Sanjay, Ganorkar, Ashok and Ghatol, 2007 [20]	—	—	—	—
Anil, Ross and Prabhakar, 2004 [22]	• Immigration and naturalization service accelerated service system (INSPASS) • Border passage system using iris recognition at London's Heathrow airport. • The Face Pass system from Viisage is used in POS verification applications like ATMs	• Ensure that the rendered services are accessed only by a legitimate user and no one else.	—	—
Bramhananda, Reddy and Goutham, 2018 [12]	—	—	—	—

Reference	Application	Reason of application	Advantages	Disadvantages
Roy and Bandyopadhyay, 2017 [1]	—	—	—	—
Phadke, 2013 [13]	• ID management solution controlled and operated by governments	—	• Biometric identification can provide extremely accurate, secured access to information; fingerprints, retinal and iris scans produce absolutely unique datasets when done properly. • Iris scanning is less intrusive than retinal recognition because the iris is easily visible from several feet away.	—
Rui and Yan, 2018 [24]	—	—	—	—
Jin-Hyuk, Eun-Kyung and Sung-Bae, 2004 [7]	—	—	—	—
Manisha and Kumar, 2019 [8]	—	—	—	—
Himanshu, 2013 [9]	• National border controls as living passport. • Computer login • Secure access to bank account at ATM machine • Ticketless travel • Authentication in networking • Permission access control to home, office, laboratory, etc. • Driving licenses, and other personal certificates	—	—	—
Williams, 1997 [25]	—	—	—	—
Wildes, 1997 [15]	—	—	—	—
Wildes, Asmuth, Green, Hsu, Kolczynski, Matey and	—	—	—	—

Reference	Application	Reason of application	Advantages	Disadvantages
McBride, 1994 [14]				
Ganorkar and Ghatol, 2007 [19]	—	—	—	—
Daugman, 1993 [4]	—	—	—	—
Lim, Lee, Byeon and Kim, 2001 [21]	—	—	—	—

Table 5.
An overview of literature for their applications and advantageous features.

system" [4, 17]. There are similar reports for various kinds of applications and methodologies. **Table 5** describes the implemented application type and the reason for using it by mentioning the advantages and disadvantages of the proposed methods.

3. Conclusion

Biometrics means the automatic identification of a person based on his behavioral and/or physiological unique characteristics. Iris biometrics is an efficient, safe, cost-effective, easy-to-use technique for identity verification. This study provides detailed information related to iris recognition techniques. Several author's works, related to iris recognition technology, are discussed, compared and analyzed. A detailed analysis of various studies is made. Various methods are taken into account to extract features of the iris such as wavelet beam analysis and static measurement feature transformation. The main focus is on iris as biometrics feature for the secure authentication and uniqueness of human identification around the world. The iris is one of the biophysiological features that are very reliable in identification systems. It is used in multimodal biometrics and in conjunction with cryptography. It is also considered one of the fairest biometrics of the face. However, it has been found that the localization of the iris is affected by tissue. When not properly interpreted, commercial iris-based biometrics systems provide inaccurate results while identifying humans. Moreover, it is important that the iris-based identification systems work with both ideal and imperfect iris images, otherwise safety will be at stake.

As a future work, there is a scope to improve the problems related to iris recognition, specially, the issues related to the capturing Iris by the sensors. One of the innovations is the touchless Iris sensors, which will be sufficient for various difficult situations including COVID-19 in the current time and age. It will decree the need to touch the devices. This technique is needed to show its reliability and efficacy as an alternative to regular sensors. Relying on an iris recognition in a different government domain is also recommended. Implementing iris recognition technology is not only useful for Government, but other organizations and communities can also think and may benefit by applying iris recognition techniques to identify and verify.

It has also been noted that iris-based biometric systems tend to present erroneous results in uncooperative settings. Another important idea is that the iris can be used for mobile phone communications with smart devices. Revocable biometrics is useful for strong security in the event of attacks. There are direct and indirect attacks on multimodal biometrics that must be overcome. More research is needed to know that attacks like these cannot break the security of biometric systems. With these ideas in mind, in the future, people can focus on designing ATMs with iris recognition in the banking industry.

There is also a need of the time to concentrate on using real apps to support the generation of tiny iris codes for cell phones and PDAs. In this chapter, an attempt is made to provide an insight into different iris recognition methods. Technology survey provides a platform for developing new technologies in this field as a future work.

Author details

Muhammad Sarfraz* and Nourah Alfialy
Department of Information Science, College of Life Sciences, Kuwait University, Sabah AlSalem University City, Shadadiya, Kuwait

*Address all correspondence to: prof.m.sarfraz@gmail.com

References

[1] Roy S, Bandyopadhyay SK. Iris recognition using biometric techniques. International Journal of Applied Research 2017;**3**(5):822-823

[2] El-Abed M, Giot R, Hemery B, Rosenberger C. Evaluation of biometric systems: A study of users' acceptance and satisfaction. International Journal of Biometrics, Inderscience. 2012. pp.1-27. 10.1504/IJBM.2012.047644. hal-00984024

[3] Choudhary D, Tiwari S, Singh AK. A survey: Feature extraction methods for iris recognition. International Journal of Electronics Communication and Computer Technology. 2012;**2**(6): 275-279

[4] Daugman JG. High confidence recognition of persons by a test of statistical independence. IEEE Transaction on PAMI. 1993;**15**(11): 1148-1161

[5] Rakesh T, Khogare MG. Survey of biometric recognition system for Iris. International Journal of Emerging Technology and Advanced Engineering. 2012;**2**(6):272-276

[6] Bowyer KW, Hollingsworth K, Flynn PJ. Image understanding for iris biometrics, a survey. Computer Vision and Image Understanding. 2008;**110**(2): 281-307

[7] Jin-Hyuk H, Eun-Kyung Y, Sung-Bae C. A review of performance evaluation for biometrics systems. International Journal of Image and Graphics. 2005;**5** (3):501-536

[8] Manisha KN. Cancelable Biometrics: A comprehensive survey 2020;**53**:3403–3446. DOI: 10.1007/s10462-019-09767-8

[9] Himanshu S. Personal identification using iris recognition system, A Review. International Journal of Engineering and Applications (IJERA). 2013;**3**:449-453

[10] Daugman J, Dowing C. Epigenetic randomness, complexity, and singularity of human iris patterns. Biological Sciences. 2001:1737-1740

[11] Daugman J. How Iris recognition works. IEEE Transaction on Circuits and Systems for Video Technology. 2004; **14**(1):21-30

[12] Bramhananda Reddy MV, Goutham V. Iris Technology: A review on iris based biometric systems for unique human identification. International Journal of Research—Granthaalayah. 2018;**6**(1):80-90. DOI: 10.5281/zenodo.1162210

[13] Phadke S. The importance of a biometric authentication system. The SIJ Transactions on Computer Science Engineering & its Applications (CSEA). 2013;**1**(4):128-132

[14] Wildes R, Asmuth J, Green G, Hsu S, Kolczynski R, Matey J, et al. A system for automated iris recognition. Sarasota, FL: Proceedings IEEE Workshop on Applications of Computer Vision; 1994. pp. 121-128

[15] Wildes R. Iris recognition: An emerging biometric technology. Proceedings of the IEEE. 1997;**85**(9): 1348-1363

[16] Sabhanayagam T, Prasanna VV, Senthamaraikannan K. A comprehensive survey on various biometric systems. International Journal of Applied Engineering Research. 2018;**13**(5): 2276-2297

[17] Daugman. Recognizing persons by their Iris patterns. In: Biometrics: Personal Identification in Networked Society. Kluwer; 1998. pp. 103-121

[18] Sheela SV, Vijaya PA. Iris recognition methods survey. IJCA. 2010; **3**(5):19-25

[19] Ganorkar SR, Ghatol AA. Iris recognition: An emerging biometric technology. In: Proceedings of international Conference ICSCI 2007. Hyderabad; 2007. pp. 596-600

[20] Sanjay R, Ganorkar AA, Ghatol. Iris recognition, an emerging biometric technology. In: Proceedings of 6th WSEAS International Conference on Signal Processing, Robotics and Automation, Greece. 2007. pp. 91-96

[21] Lim S, Lee K, Byeon O, Kim T. Efficient iris recognition through improvement of feature vector and classifier. ETRI Journal. 2001;**23**(2): 61-70

[22] Anil JK, Ross A, Prabhakar S. An introduction to biometric recognition. IEEE Transactions on Circuits and Systems for Video Technology. 2004; **14**(1):4-20

[23] Arrawatia S, Kishore B, Mitra P. Critical literature survey on iris biometric recognition. International Journal of Scientific Research in Science and Technology IJSRST. 2017;**3**(6): 600-605

[24] Rui Z, Yan Z. A survey on biometric authentication: Toward secure and privacy-preserving identification. IEEE Access. December 2018;**99**:1-1. DOI: 10.1109/ACCESS.2018.2889996

[25] Williams G. IRIS recognition technology. IEEE AES System Magazine. 1997:23-29

Chapter 2

Biometric-Based Human Recognition Systems: An Overview

David Palma and Pier Luca Montessoro

Abstract

With the proliferation of automated systems for reliable and highly secure human authentication and identification, the importance of technological solutions in biometrics is growing along with security awareness. Indeed, conventional authentication methodologies, consisting of knowledge-based systems that make use of something you know (e.g., username and password) and token-based systems that make use of something you have (e.g., identification card), are not able to meet the strict requirements of reliable security applications. Conversely, biometric systems make use of behavioral (extrinsic) and/or physiological (intrinsic) human characteristics, overcoming the security issues affecting the conventional methods for personal authentication. This book chapter provides an overview of the most commonly used biometric traits along with their properties, the various biometric system operating modalities as well as various security aspects related to these systems. In particular, it will be discussed the different stages involved in a biometric recognition process and further discuss various threats that can be exploited to compromise the security of a biometric system. Finally, in order to evaluate the systems' performance, metrics must be adopted. The most widely used metrics are, therefore, discussed in relation to the provided system accuracy and security, and applicability in real-world deployments.

Keywords: biometrics, authentication, identification, human traits, evaluation criteria, pattern recognition system, security, vulnerabilities

1. Introduction

This chapter stands as an introduction to the field of biometrics which is rising as an advanced layer to many user- and enterprise-centric security systems. In fact, conventional authentication methods, such as traditional passwords, have long been a weak point for security systems. Biometrics aims to answer this issue by linking proof-of-identity to our physiological traits and behavioral patterns. It is therefore important to present the concepts and primitives of performance metrics due to their impact on secure biometric systems. Thus, a brief overview is given to describe the main biometric traits along with their properties as well as the various biometric system operating modalities and the relatively known vulnerabilities. Finally, the criteria for performance evaluation have been defined to determine the system accuracy and security which are related to the applicability in real-world deployments.

2. Biometric traits

Various biometric modalities have been developed over the years making the biometric technology landscape very vibrant. Prominent examples of physiological/biological and behavioral biometric characteristics, which have been the purpose of major real-world applications, are illustrated in **Figure 1**.

2.1 Physiological/biological (intrinsic) human characteristics

Biological biometrics make use of traits at a genetic and molecular level which may include features like DNA or blood, whilst physiological biometrics involve the individual physical traits like a fingerprint, iris, or the shape of the face. On the other hand, behavioral biometrics are based on patterns unique to each person, for example, how an individual walks, speaks, or even types on a keyboard. Some examples of biometric traits are briefly described below.

Fingerprint: Fingerprint recognition, which measures a finger's unique pattern, is one of the oldest forms of biometric identification. This trait appears as a series of dark lines and white spaces when captured from the device and it consists of a set of ridges and valleys located on the surface tips of a human finger to uniquely distinguish individuals from each other. The fingerprint features are generally categorized into—(i) macroscopic ridge flow patterns (core and delta points), (ii) minutia features (which consists of the ridge bifurcations/trifurcation and the ridge endings), and (iii) pores and ridge contour attributes (incipient ridges, pore, shape, and width). Fingerprints of identical twins are different and so are the prints on each finger of the same person [1].

Face: Facial features use the location and shape (geometry) of the face, including the distance between the eyes, the distance from the chin to the forehead, or other measures that involve eyebrows, nose, lips, and jawline [2]. This kind of recognition is

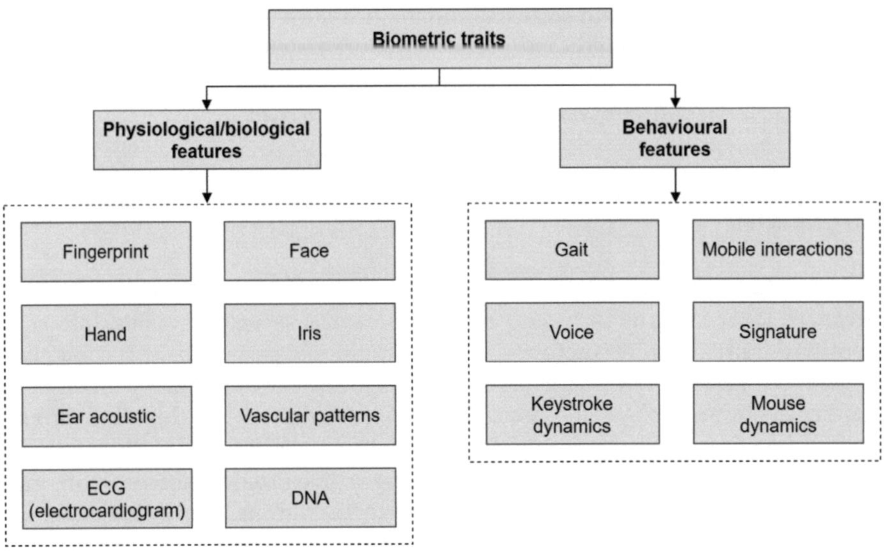

Figure 1.
Examples of physiological/biological and behavioral traits applied in biometric recognition applications.

a nonintrusive method with reasonable authentication performance in commercially available systems. However, several constraints may be imposed by the systems on how the facial images are obtained to work properly, for example, controlled illumination and background. Moreover, its susceptibility to change due to factors such as aging or expression may present a challenge [3].

Hand geometry: This trait is based on the geometric characteristics of the hand such as the length and width of fingers, their curvature, and their relative position to other features of the hand. Though once a dominant method of biometric measurement due to the requirement of the low complexity in feature extraction and low-cost imaging, modern advances in biometrics have replaced its relevance in most applications [4]. Furthermore, such a biometric trait is not known to be very distinctive and hand geometry-based recognition systems cannot be scaled up for systems requiring the identification of an individual from a large population. In addition, hand-geometry features from both hands are expected to be similar, as their anatomy is quite similar [5].

Iris: Systems based on this trait are among the most accurate biometric systems available. This human characteristic refers to the colored part in the eye that consists of thick, thread-like muscles characterized by unique folds and patterns that can be used to identify and verify the identity of humans. Furthermore, this biometric trait is stable because iris patterns do not vary during the course of a person's life and are not susceptible to loss, manipulation, or theft, making an iris recognition system robust to spoofing attacks. One interesting point worth noting is that even the two eyes in the same person have different patterns [6].

Ear acoustic: The main purpose of this kind of recognition system is to map one aspect within acoustic ear recognition, namely the performance of the ear characteristics bands and peaks. An ear signature is generated by probing the ear with inaudible sound waves which are reflected bouncing in different directions and picked up by a small microphone. The shape of the ear canal determines the acoustic transfer function which forms the basis of the signature. The recognition process is also possible, whilst the subject is on the move and caters to the protection of secrecy, which expands the applicability of this technology [7].

Vascular patterns: This biometric trait has been largely investigated for its advantages over other features. In fact, the vascular pattern of the human body is unique to every individual, even between identical twins [8], remains steady during the course of a person's life, and lies underneath the human skin ensuring confidentiality and robustness to counterfeiting, as opposed to other intrinsic and extrinsic biometric traits that are more vulnerable to spoofing, thus leading to important security and privacy concerns [9]. To acquire the network structure of blood vessels underneath the human skin, a vascular-based recognition system uses near-infrared light to reflect or transmit images of blood vessels, since they are almost invisible in normal lighting conditions [10]. The most commonly used vascular biometric solutions use hand-oriented modalities, such as finger vein, palm vein, hand dorsal vein, and wrist vein recognition, as well as eye-oriented modalities, such as retina and sclera recognition [11].

Electrocardiogram (ECG): This trait considers the human heart and body anatomic features form the shape of the ECG signal typically acquired using a few electrodes, amplifiers, filters, and a data acquisition module, and which reports the strength and timing of the electrical activity of the heart [12]. However, scientific findings to date throw doubt on the specificities of real-world application scenarios and acceptability by the potential end users, which pose several constraints and questions.

Deoxyribonucleic acid (DNA): DNA matching is based on a common molecular biology method named short tandem repeat (STR)[1] analysis, which is used to compare allele repeats at specific locations on a chromosome in DNA between two or more samples [14, 15]. DNA-based biometric recognition has been widely used in forensic science and scientific investigation due to its very high accuracy, despite the fact that identifications require tangible physical samples and cannot be done in real time.

2.2 Behavioral (extrinsic) human characteristics

Keystrokes, handwriting, gait, how a person uses a mouse, and other movements are some of the behavioral traits that a biometric system may analyze to assess the individual's identity.

Gait: This characteristic may be changeable over a large time span due to various reasons, such as weight gain [16]. Thus, it can be used in low-security applications for massive crowd surveillance as it can quickly identify people from afar based on their walking style, even harnessing the potential of a large number of surveillance cameras installed in public locations into a biometric system. In fact, such a system does not require the individuals to be cooperative, nor that they wear any special device or equipment to be recognized [17].

Mobile interactions: It is based on the unique ways in which users swipe, tap, pinch-zoom, type, or apply pressure on the touchscreen of mobile devices like tablets and phones, thus providing characteristic patterns that may be used to identify people, even considering further features deriving from on-board sensors such as GPS, gyroscope, and accelerometers [18], which can also be configured to collect data in passive mode. Therefore, mobile interactions-based biometrics focuses not so much on the outcome of the user's actions but rather on the way a user performs those actions.

Signature: Signature recognition is the most widely accepted method for documents authentication and it makes use of shorter handwriting probes compared to text-independent writer recognition methods, but it requires to write the same sign every time. A signature authentication scheme can be categorized into two methods—(i) off-line or static (the signature is digitized after the writing process) and (ii) online or dynamic (the signature is digitized during the writing process). Signature biometric features are extracted by analyzing curves, edges, spatial coordinates, inclination, the center of gravity, pen pressure, and pen stroke of the signature samples in both off-line and online applications. However, dynamic information like writing speed and stroke order is available only in online signatures [19].

Mouse dynamics: It makes use of patterns in mouse or trackpad cursor movement including clicks, trajectories, direction changes, tracking speed, and the relationships between them. Mouse-generated movement features are relatively stable for the same individual and different compared to other users, as such can be used to authenticate individuals [20]. These methods are most often used to continuously verify the user's identity.

Keystrokes: Keystroke dynamics (also known as typing biometrics) include the tracking of the rhythm used to type on a keyboard. Two events constitute a keystroke event—key down and key up. The first one occurs when an individual presses a key, whilst the second one is associated with the event that occurs when the pressed key is

[1] STR is the DNA sequence of the short repeat region of the sequence in the noncoding region of the human genome [13].

released. Making use of these events, a set of inter-key and intra-key features known as delay times, hold times, and key down-key downtimes can be extracted. In general, keystroke recognition will work on the computer or virtual keyboards, mobile phones, smartwatches, and touchscreen panels, providing a low-cost authentication method that can be easily deployed in a variety of scenarios [21].

Voice: Voice recognition technology falls under both the physiological and behavioral biometric categories. Voice biometric recognition allows to distinguish among humans' voice for personal authentication as voice features include physical characteristics such as vocal tracts, nasal cavities, mouth, and larynx [22]. Behaviorally, the way a person speaks or says something, for example, tone, movement variations, accent, pace, and so on, is also considered unique to each individual. Using data from both physiological and behavioral biometrics creates, therefore, a precise vocal signature, though mismatches may occur due to illness or other factors.

2.3 Properties of biometric traits

The main requirements that should be satisfied before a trait can be characterized as suitable for its applicability in a biometric recognition system, are briefly discussed as follows [23].

- Universality: Every individual or at least most of them, accessing the biometric application should possess the characteristic.

- Distinctiveness (or uniqueness): The given trait should be sufficiently different across individuals comprising the user population. Otherwise, the proportion of times the biometric system grants access to unauthorized individuals would be unacceptably high.

- Permanence: The biometric trait of an individual should be sufficiently invariant (with respect to the matching criterion) over a period of time. This implies that the given trait should not change significantly over time otherwise the proportion of times the biometric system denies access to authorized individuals would be unacceptably high.

- Collectability: The biometric trait can be measured quantitatively with particular regard to the easiness of obtaining the biometric data using suitable devices that do not cause undue inconvenience to the user.

Even though any human characteristic can be used as a biometric trait as long as the previous requirements are satisfied, in real-world biometric recognition applications there are a number of other issues that should be considered, such as:

- Performance: This is a property aimed at assessing the verification or identification accuracy, the computational time required for a single recognition, as well as the operational and environmental factors that may affect or not the recognition accuracy and speed.

- Acceptability: It indicates the extent to which people are willing to accept the use of a specific biometric application as well as their willingness to provide their biometric data. Nowadays, this is a crucial aspect to be considered due to the current pandemic

Biometric trait	Universality	Uniqueness	Permanence	Collectability	Performance	Acceptability	Circumvention
Fingerprint	M	H	H	M	H	M	H
Face	H	L	M	H	L	H	H
Hand geometry	M	M	M	H	M	M	M
Iris	H	H	H	M	H	L	L
Ear	M	M	H	M	M	H	M
Vascular patterns	H	H	M	M	H	M	L
DNA	H	H	H	L	H	L	L
Gait	M	L	L	H	L	H	M
Signature	L	L	L	H	L	H	H
Keystroke dynamics	L	L	L	M	L	M	M
Voice	M	L	L	M	L	H	H

H = High; M = Medium; L = Low.

Table 1.
Comparison study of the most common traits based on the characteristics of biometric entities.

situation caused by the severe acute respiratory syndrome coronavirus 2 (SARS-CoV-2) [24], raising questions about how safe using touch-based biometric systems really is as touching the sensors can potentially spread viruses. As a consequence, less-constrained biometrics will likely be the preferred modality, whilst there may be less demand for other solutions that rely on physical contact with a reader.

- Circumvention: This property reflects how easily the system can be deceived through potential spoofing attacks. It refers to the ways in which an attacker can endeavor to bypass a biometric system and finally attack the weak spot of such a system in order to gain unauthorized access.

Real-life biometric recognition systems ought to meet the requirements of accuracy, speed, and resource constraints, be harmless to the users, be accepted by the intended population as well as sufficiently robust to various fraudulent methods and attacks to the system [25].

Table 1 is reported a comparison study of the most popular traits based on the characteristics of biometric entities [26].

3. Biometric system operating modes

A biometric system can provide two kinds of operating modes (identity management functionalities), namely, *verification* and *identification*. Biometric systems can indeed automatically authenticate[2] or identify subjects in a reliable and fast way and

[2] Throughout this book chapter, the term authentication will be used as a synonym for verification.

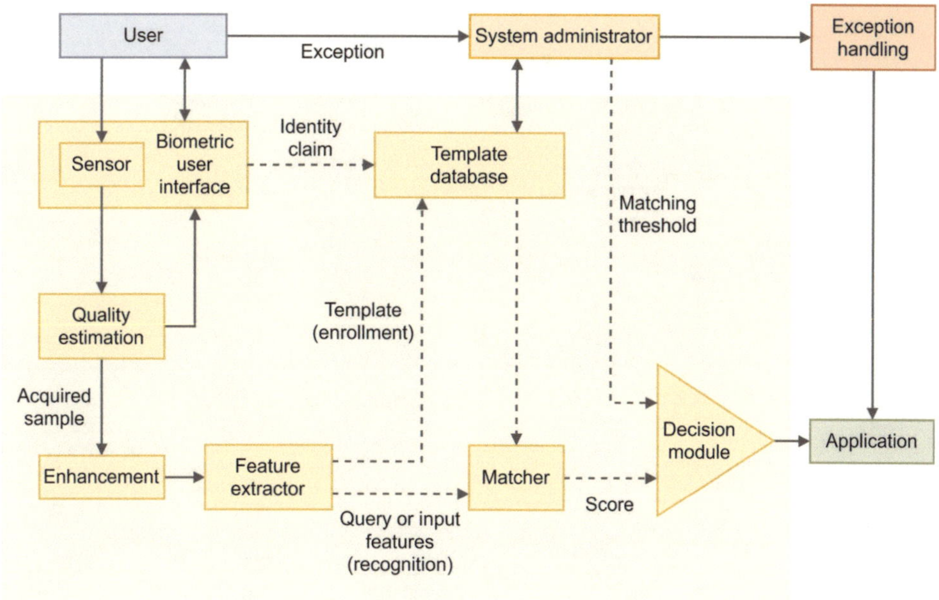

Figure 2.
Basic building blocks of a generic biometric recognition system.

are, therefore, suitable to be used in a wide range of applications to face the risks of unauthorized logical or physical access and identity theft, as well as new threats such as terrorism or cybercrime [27]. **Figure 2** provides a high-level view of a generic biometric recognition system as well as all its basic building blocks, whilst **Figure 3** depicts the enrollment and the biometric recognition schemes of the authentication and identification modalities.

3.1 Authentication

In the authentication mode, the purpose of the biometric system is to verify whether an individual's claimed identity is genuine or not (binary classification). Thus, the captured biometric data (query) is compared only with the biometric template(s) stored in the system database and corresponding to the claimed identity (one-to-one or one-to-few comparison). Given a claimed identity I and a query feature set $\boldsymbol{x}^Q$, the biometric system has to be categorized $(I, \boldsymbol{x}^Q)$ into "genuine" or "impostor" class. Let $\boldsymbol{x}_I^E$ be the stored biometric template corresponding to the identity I (i.e., the enrolled user with identity I). The similarity measure between $\boldsymbol{x}^Q$ and $\boldsymbol{x}_I^E$ gives, as a result, a matching score. Hence, the biometric system applies the decision rule given by

$$(I, \boldsymbol{x}^Q) \in \begin{cases} \text{genuine}, & \text{if } s(\boldsymbol{x}^Q, \boldsymbol{x}_I^E) \geq \xi, \\ \text{impostor}, & \text{otherwise}, \end{cases} \tag{1}$$

where s represents a similarity function and ξ represents a pre-defined threshold at which the system is intended to operate. The authentication mode is typically employed for positive recognition, where the aim is to prevent multiple people from using the same identity [28].

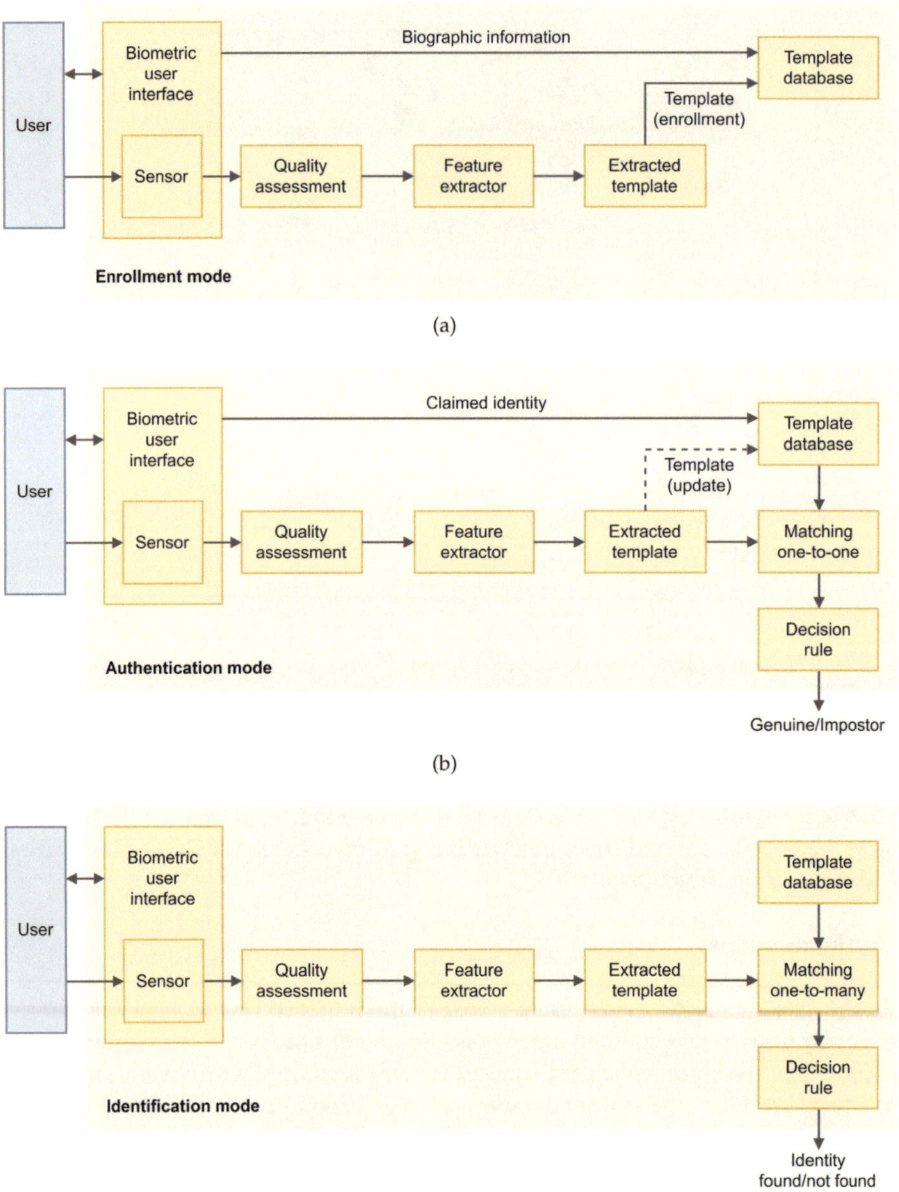

Figure 3.
Different operating modes of a biometric system—(a) enrollment mode, (b) authentication mode (the dashed line is an optional operation aimed at updating a specific user's template), and (c) identification mode.

3.2 Identification

In the identification mode, the purpose of the biometric system is to recognize an individual's identity by searching the templates of all the enrolled individuals in the system database for a match (one-to-many comparison) without the subject having to claim an identity.

DOI: http://dx.doi.org/10.5772/intechopen.101686

This operating mode can be further split into negative and positive identification—in the negative identification (also known as *screening*), the user is considered to be hiding her/his true identity from the biometric system, whilst in the positive identification, the user tries to positively identify herself/himself to the system without explicitly claiming an identity. Given a query feature set $\boldsymbol{x}^Q$, the biometric system has to determine the identity $I_k \; \forall k \in \{1, 2, \dots, n, n+1\}$ where $\{I_1, I_2, \dots, I_n\}$ are identities of the enrolled users in the system, whilst I_{n+1} represents the failure case where no identity can be assigned for the given query (*open-set identification*). Hence, assuming that $\boldsymbol{x}^E_{I_k}$ is the stored template corresponding to the identity I_k, the biometric system applies the decision rule given by

$$\boldsymbol{x}^Q \in \begin{cases} I_k, & \text{if} \max_k \left\{ s\left(\boldsymbol{x}^Q \boldsymbol{x}^E_{I_k}\right)\right\} \geq \xi, \\ I_{n+1}, & \text{otherwise}, \end{cases} \quad (2)$$

where s represents a similarity function and ξ represents a pre-defined threshold at which the system is intended to operate.

The identification mode is typically employed for screening[3], where the aim is to prevent a single person from using multiple identities [28].

4. Vulnerabilities

Biometric-based cybersecurity solutions ensuring tight access control are essential in preventing intrusions and unauthorized accesses. However, even though a biometric system enhances user convenience and security, does not necessarily mean that it is also exempt from security and privacy issues. Many security measures in biometric systems are designed to protect one or more facets of the CIA triad, which is a common framework that refers to confidentiality, integrity, and availability [31].

- Confidentiality is roughly equivalent to privacy. Measures undertaken to ensure confidentiality are designed to prevent sensitive information from reaching unauthorized people. It is perhaps the most obvious aspect of the CIA triad when it comes to security; but correspondingly, it is also the one which is attacked most often. Confidentiality covers a wide spectrum of access controls and measures that protect data from getting misused by any unauthorized access. Cryptography and encryption methods are an example of an attempt to prevent illegitimate access ensuring the confidentiality of (sensitive) data.

- Integrity of information refers to the ability to protect information from being modified or destroyed by unauthorized parties, thus ensuring nonrepudiation and authenticity of the information. Thus, integrity involves maintaining the consistency and trustworthiness of data. One type of security attack is to intercept some important data and make changes to it before sending it on to the intended receiver.

---

[3] In some real scenario, such as latent palmprint matching [29], it is preferable to use a semi-automated approach aimed at providing the top n identities that best match to the given template for further analysis by a human expert. Alternatively, it is possible to consider all the identities whose corresponding match scores exceed the threshold ξ that leads to a challenging task in a quite large database (e.g., FBI's next generation identification (NGI) system, which provides the world's largest repository of biometric and criminal history information [30]).

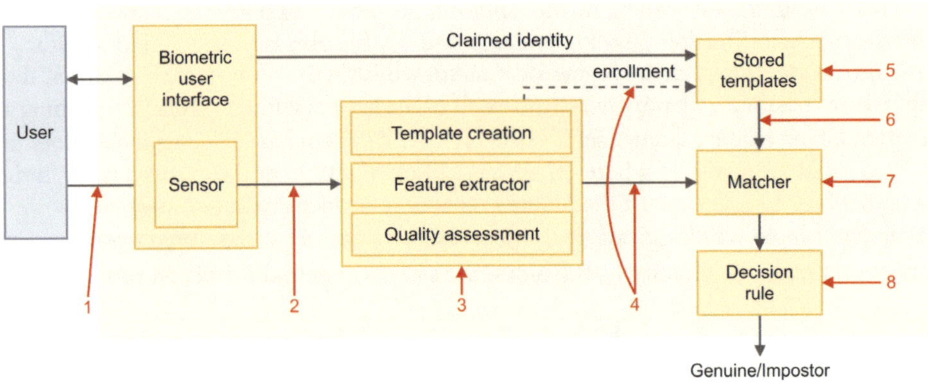

Figure 4.
Attack points of a general biometric system.

- Availability of information refers to ensuring that only legitimate and authorized parties are able to access the information when needed. Problems affecting the information system could make it impossible to access information, thereby making the information unavailable. Some types of security attacks attempt to deny access to the appropriate user, either for the sake of inconveniencing them, or because there is some secondary effect.

Biometric recognition systems implicitly (and effectively) address the authentication problem included in the last issue of the CIA triad, which consists in guaranteeing access to data only to authorized users. The reason for this is because biometric traits are (generally) not susceptible to loss, manipulation, or theft, and therefore overcome the security issues affecting the conventional methods for personal authentication, such as knowledge-based and token-based systems. However, it must be kept in mind that a biometric-based security solution is composed of several different components and the recognition module, which is only capable of addressing the authentication aspect, is just one of them. Thus, a logical structure-based approach of biometric systems is used to describe the eight points of attacks illustrated in **Figure 4**.

1. An attack on the biometric sensor consists of presenting a fake biometric trait (e.g., an artificial characteristic) to perform a spoofing attack aimed to either avoid detection (false negative) or masquerade as another (false positive). Methods used to prevent spoofing attacks include layered biometrics, liveness, and combining biometrics and conventional authentication methods such as passwords, tokens, or smart cards [32].

2. The connection between the biometric sensor and the subsequent modules of the system may be attacked to allow input of a stored digital biometric signal. This data can be obtained, for instance, by performing an eavesdropping (disclosure) attack [31].

3. Attacks on the feature extractor can be used either to create impostors or to evade detection. Hence, knowledge of the algorithms involved in this module[4]

[4] Since biometric recognition algorithms are likely susceptible to reverse engineering techniques, it is possible to conduct off-line experiments on a copy of the biometric software to be hacked in order to achieve the objective [32].

may be used to forge features in presented samples to cause computation of incorrect features. To achieve this, an attacker can replace the feature extractor with a Trojan horse program that produces the desired feature sets.

4. An attack on the output of the previous module consists of spoofing the legitimate biometric feature set to replace it with a synthetic one.

5. Vulnerabilities of template database concern modifying the storage (modifying, removing, or adding templates), copying stored data for future use (identity theft or directly using the acquired information to gain access), or modifying the identity to which the biometric is assigned.

6. The channel between the template database and the matching module is similarly vulnerable to the previous one, however, the attack against data transmission may be easier than against the template storage, especially in the case of an adversary able to intercept any information communicated by the system by observing the data (passive eavesdropping). Encryption is crucial in this case, but may still be vulnerable to key discovery [33].

7. The matcher module is responsible for computing a similarity score between two biometric templates in order to confer the likelihood that they are from the same subject. Even though it may not be possible to do it easily, an attack against the matcher can be possible in specific cases. For instance, it is possible to replace the matcher module with a Trojan horse program that always outputs high scores thereby defying system security [34].

8. An attack on the final decision module means that if the final decision can be inserted or blocked by the attacker then the authentication system function will be overridden. If it is instead reviewed by a human operator, a DoS (denial of service) attack may be performed to mislead it or to force it to mistrust the output of the system [35].

5. Criteria for performance evaluation

The reliability and validity of a biometric scheme as well as the selection of a certain biometric trait for an application are determined by specific measures that are used to evaluate the recognition accuracy and effectiveness as addressed in ISO/IEC Standards [36]. Accordingly, to evaluate the accuracy of the proposed method based on a single-sample approach for unimodal biometric systems, each sample in the database should undergo a one-to-one matching test against every single stored sample. Hence, a comparison between a subject with a real identity I_r and a subject with claimed identity I_c is aimed at testing the hypothesis:

$$H_0 : \{I_r = I_c\} \text{ versus } H_1 : \{I_r \neq I_c\} \tag{3}$$

where H_0 is the null hypothesis that the user is who s/he claims to be (genuine or intra-class matching), whilst H_1 is the alternative hypothesis that the user is not who s/he claims to be (impostor or inter-class matching). To test the hypothesis in (3), it is required to compute a similarity measure, $s(Q, T)$ where large (respectively, small) values of s indicate that the template T of the claimed identity I_c in the database and the biometric query Q of a real user I_r are close to (far from) each other. Formally, the verification problem consists of determining if a claimed identity I with biometric data Q belongs to the class H_0 or not:

$$(I, Q) = \begin{cases} H_0, & \text{if } s(Q, T) \geq \xi, \\ H_1, & \text{otherwise.} \end{cases} \tag{4}$$

Precisely, given a threshold ξ, all matching values s lower (respectively, greater) than ξ lead to the rejection (acceptance) of the null hypothesis [37]. Therefore, whether the hypothesis is accepted or not, the test is prone to two kinds of error:

- false acceptance rate (FAR), that is the probability of accepting the null hypothesis H_0 when input is not valid (type-I error),
- false rejection rate (FRR), that is the probability of rejecting the null hypothesis H_0 when input is valid (type-II error).

Let H_0 and H_1 be the labels that denote the genuine and impostor classes, respectively. Assume also that the $p(s|H_0)$ and $p(s|H_1)$ represent the probability density functions of the genuine and impostor scores, respectively. Then the FAR and FRR distributions are given by:

$$FAR(\xi) = p(s \geq \xi | H_1) = \int_{\xi}^{+\infty} p(s|H_1) ds, \tag{5}$$

$$FRR(\xi) = p(s < \xi | H_0) = \int_{-\infty}^{\xi} p(s|H_0) ds. \tag{6}$$

The false acceptance and false rejection rates are functions of the system threshold ξ and are closely related because the increase of one implies the decrease of the other. Hence, for a given biometric system, it is not possible to decrease both these errors at the same time by varying the threshold ξ [25]. The separation between the two distributions (or classes) indicates the ability of the system to distinguish the genuine user samples from those of the impostors. Indeed, the separation also provides a hint on the threshold point that maximizes the variance between the two classes in order to correctly mark a user sample image as authentic or impostor [23].

The genuine acceptance rate (GAR) is instead the probability of accepting the null hypothesis H_0 when input is valid, hence it can be used as an alternative to FRR:

$$GAR(\xi) = p(s \geq \xi | H_0) = 1 - FRR(\xi). \tag{7}$$

Depending on the security level required by the final application (i.e., forensics, surveillance and homeland security, civilian, or high-security applications), the same biometric system may operate at different threshold values (ξ), as illustrated in **Figure 5**.

Hence, in order to evaluate the biometric system performance as a function of the threshold ξ, the following curves can be considered:

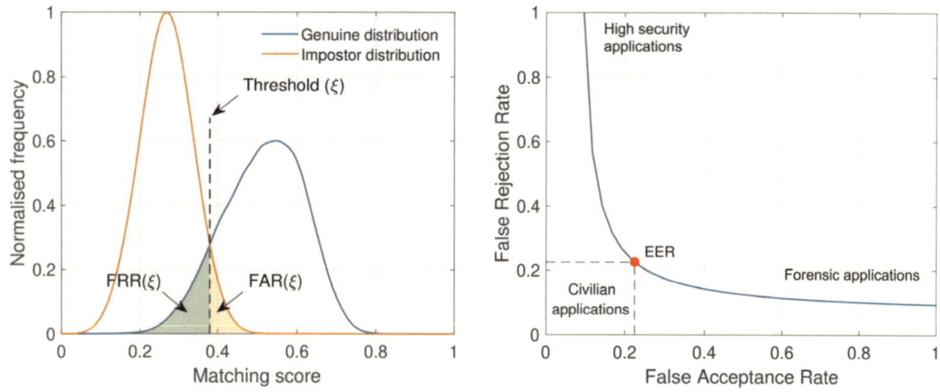

(a) (b)

Figure 5.
Examples of biometric system error rates: (a) FAR and FRR for a given threshold ξ are displayed over the genuine and impostor score distributions and (b) typical operating points of different biometric applications are displayed on a DET curve aimed at relating FAR and FRR at different threshold values.

- The receiver operating characteristic (ROC) is a graphical plot that illustrates the trade-off between false acceptance and false rejection rates when the threshold varies, whilst the intersection point for which rejection and acceptance errors are equal is named equal error rate (EER). The curve is generated by plotting the genuine acceptance rate against the false acceptance rate at various threshold settings,

- The detection error trade-off (DET) is another graphical plot that illustrates the false rejection rate against the false acceptance rate at various threshold values. The two axes are scaled nonlinearly by their standard normal deviates[5] or just by logarithmic transformation.

Furthermore, the above-mentioned ROC and DET curves are threshold-independent, allowing performance comparison of different biometric systems under similar conditions [23], as illustrated in **Figure 6**. Given a set of thresholds $\{\xi_i\} \mid s_{\min} \le \xi_i \le s_{\max} \ \forall i \in \{1, 2, \dots, n\}$ where $s_{\min}$ and $s_{\max}$ are the minimum and maximum scores, respectively, in a given set of match scores $\{s_i\} \mid 0 \le s_i \le 1 \ \forall i \in \{1, 2, \dots, n\}$. Then, it is possible to generate a ROC curve computing the overall false acceptance and false rejection rates for each threshold value ξ as follows:

$$FAR = \frac{1}{N}\sum_{k=1}^{N} FAR(\xi), \tag{8}$$

$$FRR = \frac{1}{N}\sum_{k=1}^{N} FRR(\xi), \tag{9}$$

---

5 In the normal deviate scale, the threshold values ξ correspond to linear multiples of standard deviation σ of a Gaussian distribution. Thus, if the FAR and FRR distributions are Gaussian, the corresponding DET curve would be linear [25].

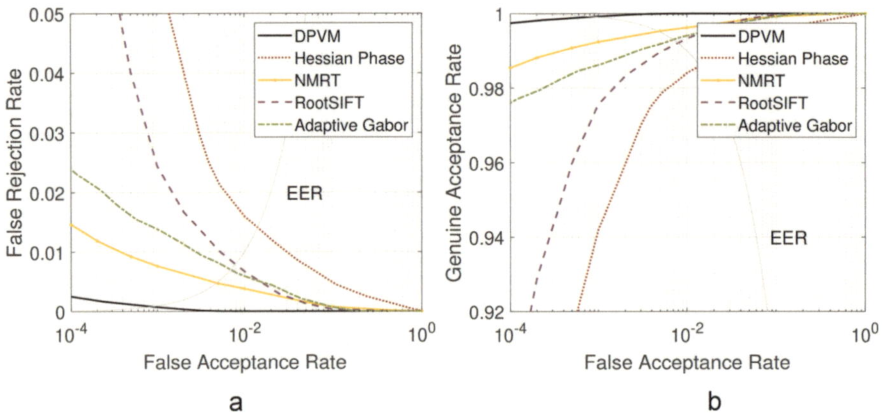

Figure 6.
Example of vascular-based biometric systems performance comparison [4]. Comparative graph of—(a) DET curves generated by plotting FRR against FAR and (b) ROC curves generated by plotting GAR against FAR.

where N represents all identities being evaluated by the system and

$$FAR(\xi) = \frac{\text{no.of FARs}}{\text{no.of impostor accesses}} \tag{10}$$

$$FRR(\xi) = \frac{\text{no.of FRRs}}{\text{no.of genuine accesses}}. \tag{11}$$

Since biometric systems cannot jointly provide a false acceptance rate equal to zero and a perfect verification/identification rate, the system threshold must be adjusted for the given application considering the trade-off between accuracy and false positives. Once the threshold has been set, the system can be evaluated by means of common measures that are used to assess the classification accuracy and effectiveness. In this context, we are interested in confirming or denying the identity of a subject leading thus to a dichotomous binary classification problem, where the labels are P (genuine) and N (impostor) and the predictions of the classifier are summarized in a 2×2 contingency table known as confusion matrix [38] (expanded in **Table 2**):

$$\mathbf{M} = \begin{bmatrix} TP & FN \\ FP & TN \end{bmatrix} \tag{12}$$

		Predicted class		
		P	***N***	**Total**
Actual class	***P***	*TP*	*FN* (Type-II error)	*TP + FN*
	N	*FP* (Type-I error)	*TN*	*FP + TN*
	Total	*TP + FP*	*FN + TN*	

Table 2.
Example of confusion matrix for a dichotomous binary classification problem.

which completely describes the outcome of the classification task. This contingency table may be expressed using raw counts of the number of records from class times each predicted label is associated with each actual class. As illustrated in **Table 2**, the confusion matrix reports:

- true positive (TP), the probability of correctly accepting the null hypothesis;
- true negative (TN), the probability of correctly rejecting the null hypothesis;
- false positive (FP), the probability of falsely rejecting the null hypothesis;
- false negative (FN), the probability of falsely accepting the null hypothesis.

Based on the entries in the confusion matrix, the total number of correct predictions carried out by the model is $TP + TN$, whilst the number of incorrect predictions is $FP + FN$ [39]. Therefore, if.

$$\boldsymbol{M} = \begin{bmatrix} n^+ & 0 \\ 0 & n^- \end{bmatrix} \tag{13}$$

where obviously $n^+ = TP + FN$ and $n^- = FP + TN$, then the classification has been perfectly done. Conversely, if the confusion matrix is as follows

$$\boldsymbol{M} = \begin{bmatrix} 0 & n^+ \\ n^- & 0 \end{bmatrix} \tag{14}$$

it represents the worst case (perfect misclassification).

Several measures have been defined to assess the quality of a prediction [40], aimed at conveying into a single figure the structure of $\boldsymbol{M}$. The most used functions are briefly described as follows.

Precision also known as positive predictive value (PPV) counts the true positives, how many samples are properly classified within the same cluster (closeness of the measurements to each other)

$$PPV = \frac{TP}{TP + FP}. \tag{15}$$

Sensitivity also known as recall or true positive rate (TPR) refers to the proportion of the samples properly classified as true positives out of the actual number of true positives

$$TPR = \frac{TP}{TP + FN}. \tag{16}$$

F-measure combines precision and recall in a single metric, indeed, it is the harmonic mean of precision and sensitivity and as a function of $\boldsymbol{M}$, has the following form:

$$F_1 = 2\frac{PPV \cdot TPR}{PPV + TPR} = \frac{TP}{TP + \frac{1}{2}(FN + FP)} \tag{17}$$

where the worst case ($F_1 = 0$) is achieved for $TP = 0$, whilst the best case ($F_1 = 1$) is reached for $FN = FP = 0$.

Accuracy represents the ratio between the correctly predicted instances and all the instances in the dataset, whose range is between 0 (worst case) and 1 (best case):

$$ACC = \frac{TP + TN}{TP + TN + FP + FN}. \tag{18}$$

Matthews correlation coefficient is the measure of the quality of binary (two-class) classifications:

$$MCC = \frac{TP \cdot TN - FP \cdot FN}{\sqrt{(TP + FP)(TP + FN)(TN + FP)(TN + FN)}} \tag{19}$$

it is a correlation coefficient between the actual and predicted binary classifications and it returns a value between −1 (worst case) and 1 (best case).

Accuracy and F-score computed on confusion matrices have been (and still are) among the most popular adopted metrics in binary classification tasks. However, these statistical measures can dangerously show overoptimistic inflated results, especially on imbalanced datasets [40]. Hence, among all the parameters described above, the Matthews correlation coefficient (MCC) is the only one that takes into account true and false positives and negatives and is generally regarded as a balanced measure that can be used even if the classes are of very different sizes [41].

6. Conclusions

Biometric-based technologies make use of unique behavioral (extrinsic) and/or physiological/biological (intrinsic) attributes to overcome the security issues affecting the conventional methods for identity authentication. Even though biometrics has been in use for decades, the advent of technology has expanded its application from primarily criminal identification to a wide range of everyday tasks, becoming a regular security process of our nowadays life. Accurate authentication or identification is fundamental to physical security, cyber security, military applications (e.g., biometric-driven lethal autonomous weapon systems), financial transactions, contracts and employment, public services, criminal justice, national security, and more. The approaches that have been proposed in literature depend on the type and the number of the underlying biometric traits, which, in general, cannot be easily transferred between people, and thereby represents a highly secure unique identifier. As a matter of fact, various biometric modalities have been developed over the years making the biometric technology landscape very vibrant. In this book chapter, we have provided an overview of the most commonly used biometric traits along with their properties, the various biometric system operating modalities as well as various limitations and weaknesses related to these systems. Indeed, biometric technologies have a number of vulnerabilities that underscore the concerns over their employment and may result in the failure of the technology to perform as anticipated. We have also discussed how the system threshold must be adjusted for the given application considering the trade-off between accuracy and false positives since biometric systems cannot jointly provide a FAR equal to zero and a perfect recognition rate. Finally, the criteria for performance evaluation have been defined to determine the system's accuracy and security which are related to the applicability in real-world deployments, even though the existing evaluation metrics are more related to the data quality than the security aspects of the overall system. However, despite the risks, biometrics provide very compelling security solutions remaining a growing way to verify identity offering tons of promise for the future of cybersecurity.

Conflict of interest

The authors declare no potential conflicts of interest with respect to the research, authorship, and/or publication of this article.

Abbreviations

DET	Detection error trade-off
DNA	Deoxyribonucleic acid
ECG	Electrocardiogram
FAR	False acceptance related
FN	False negative
FRR	False rejection rate
FP	False positive
GAR	Genuine acceptance rate
MCC	Matthews correlation coefficient
NGI	Next-generation identification
PPV	Positive predictive value
ROC	Receiver operating characteristic
SARS-CoV-2	Severe acute respiratory syndrome coronavirus 2
STR	Short tandem repeat
TN	True negative
TP	True positive
TPR	True positive rate

Author details

David Palma*† and Pier Luca Montessoro†
Polytechnic Department of Engineering and Architecture, University of Udine, Udine, Italy

*Address all correspondence to: david.palma@uniud.it

†D.P. and P.L.-M. designed the research; D.P. performed the research and wrote the paper. The results and the paper were analysed and reviewed by P.L.-M. All authors have read and agreed to the published version of the manuscript.

References

[1] Maltoni D, Maio D, Jain AK, Prabhakar S. Handbook of Fingerprint Recognition. London, UK: Springer Science & Business Media; 2009

[2] Zhao W, Rama Chellappa P, Phillips J, Rosenfeld A. Face recognition: A literature survey. ACM Computing Surveys. 2003;**35**(4):399-458

[3] Jain AK, Ross A, Prabhakar S. An introduction to biometric recognition. IEEE Transactions on Circuits and Systems for Video Technology. 2004; **14**(1):4-20

[4] Palma D, Montessoro PL, Giordano G, Blanchini F. Biometric palmprint verification: A dynamical system approach. IEEE Transactions on Systems, Man, and Cybernetics: Systems. 2019;**49**(12):2676-2687

[5] Li SZ, Jain AK. Encyclopedia of Biometrics: I-Z. Boston, MA: Springer Science & Business Media; 2015

[6] Daugman J. How iris recognition works. In: The Essential Guide to Image Processing. Amsterdam, NL: Elsevier; 2009. pp. 715-739

[7] Akkermans AHM, Kevenaar TAM, Schobben DWE. Acoustic ear recognition for person identification. In: Fourth IEEE Workshop on Automatic Identification Advanced Technologies (AutoID). New York, US: IEEE; 2005. pp. 219-223

[8] Kumar A, Hanmandlu M, Gupta HM. Online biometric authentication using hand vein patterns. In: IEEE Symposium on Computational Intelligence for Security and Defense Applications. New York, US: IEEE; 2009. pp. 1-7

[9] Palma D, Blanchini F, Giordano G, Montessoro PL. A dynamic biometric authentication algorithm for near-infrared palm vascular patterns. IEEE Access. 2020;**8**:118978-118988

[10] Zharov VP, Ferguson S, Eidt JF, Howard PC, Fink LM, Waner M. Infrared imaging of subcutaneous veins. Lasers in Surgery and Medicine: The Official Journal of the American Society for Laser Medicine and Surgery. 2004; **34**(1):56-61

[11] Uhl A. State of the art in vascular biometrics. In: Handbook of Vascular Biometrics. Cham: Springer; 2020. pp. 3-61

[12] Wübbeler G, Stavridis M, Kreiseler D, Bousseljot R-D, Elster C. Verification of humans using the electrocardiogram. Pattern Recognition Letters. 2007;**28**(10):1172-1175

[13] Hammond HA, Jin L, Zhong Y, Caskey CT, Chakraborty R. Evaluation of 13 short tandem repeat loci for use in personal identification applications. American Journal of Human Genetics. 1994;**55**(1):175

[14] Jeffreys AJ, Wilson V, Thein SL. Individual-specific 'fingerprints' of human DNA. Nature. 1985;**316**(6023): 76-79

[15] Tautz D. Hypervariability of simple sequences as a general source for polymorphic dna markers. Nucleic Acids Research. 1989;**17**(16):6463-6471

[16] Hu N, Tong H-L, Tan W-H, Yap TT-V, Chong P-F, Abdullah J. Human identification based on extracted gait features. International Journal on New Computer Architectures and Their Applications. 2011;**1**(2):358-370

[17] Mason JE, Traoré I, Woungang I. Machine Learning Techniques for Gait Biometric Recognition. New York, US: Springer; 2016

[18] Fierrez J, Pozo A, Martinez-Diaz M, Galbally J, Morales A. Benchmarking touchscreen biometrics for mobile authentication. IEEE Transactions on Information Forensics and Security. 2018;**13**(11):2720-2733

[19] Deore MR, Handore SM. A survey on offline signature recognition and verification schemes. In: International Conference on Industrial Instrumentation and Control (ICIC). New York, US: IEEE; 2015. pp. 165-169

[20] Sayed B, Traoré I, Woungang I, Obaidat MS. Biometric authentication using mouse gesture dynamics. IEEE Systems Journal. 2013;7(2):262-274

[21] Killourhy KS, Maxion RA. Comparing anomaly-detection algorithms for keystroke dynamics. In: IEEE/IFIP International Conference on Dependable Systems & Networks. New York, US: IEEE; 2009

[22] Delac K, Grgic M. A survey of biometric recognition methods. In: Proceedings Elmar-2004, 46th International Symposium on Electronics in Marine. New York, US: IEEE; 2004. pp. 184-193

[23] Palma D. A Dynamical System Approach for Pattern Recognition and Image Analysis in Biometrics and Phytopathology [PhD thesis]. Udine, IT: University of Udine; 2021

[24] Sarfraz M. Introductory chapter: On fingerprint recognition. In: Sarfraz M, editor. Biometric Systems. Rijeka: IntechOpen; 2021

[25] Jain AK, Ross A, Nandakumar K. Introduction to Biometrics. New York: Springer; 2011

[26] Dasgupta D, Roy A, Nag A, et al. Advances in User Authentication. New York, US: Springer; 2017

[27] Huang D, Tang Y, Wang Y, Chen L, Wang Y. Hand-dorsa vein recognition by matching local features of multisource keypoints. IEEE Transactions on Cybernetics. 2015;**45**(9):1823-1837

[28] Wayman JL. Fundamentals of biometric authentication technologies. International Journal of Image and Graphics. 2001;**1**(01):93-113

[29] Palma D, Montessoro PL, Giordano G, Blanchini F. A dynamic algorithm for palmprint recognition. In: 2015 IEEE Conference on Communications and Network Security (CNS). New York, US: IEEE; 2015. pp. 659-662

[30] Federal Bureau of Investigation (FBI). Next Generation Identification (NGI). Washington DC, US: 2021. Available from: https://www.fbi.gov/

[31] Palma D. Detection of Stealthy False-data Injection Attacks on Safety-Critical Cyber-Physical Systems. London, UK: Technical report, Imperial College of Science, Technology and Medicine; 2019

[32] Adler A, Schuckers SAC. Biometric Vulnerabilities: Overview. US, Boston, MA: Springer; 2009. pp. 1-11

[33] Sheldon FT, Weber JM, Yoo S-M, Pan WD. The insecurity of wireless networks. IEEE Security Privacy. 2012; **10**(4):54-61

[34] Prasad PS. Vulnerabilities of biometric system. International Journal

of Scientific & Engineering Research. 2013;**4**(6):1126-1129

[35] Ferguson N, Schneier B. Practical Cryptography. Vol. 141. New York: Wiley; 2003

[36] ISO/IEC JTC 1/SC 37 Biometrics. Information technology – biometric performance testing and reporting – part 1: Principles and framework. ISO/IEC. 2006;**1**:19795-19791

[37] Dass SC, Zhu Y, Jain AK. Validating a biometric authentication system: Sample size requirements. IEEE Transactions on Pattern Analysis and Machine Intelligence. 2006;**28**(12): 1902-1319

[38] David MW Powers. Evaluation: From precision, recall and f-measure to roc, informedness, markedness and correlation. Journal of Machine Learning Technologies. 2011;**2**(1):37-63

[39] Gan G, Ma C, Jianhong W. Data Clustering: Theory, Algorithms, and Applications. Pennsylvania, US: SIAM; 2020

[40] Chicco D, Jurman G. The advantages of the matthews correlation coefficient (MCC) over F1 score and accuracy in binary classification evaluation. BMC Genomics. 2020;**21**(1):6

[41] Boughorbel S, Jarray F, El-Anbari M. Optimal classifier for imbalanced data using matthews correlation coefficient metric. PLoS One. 2017;**12**(6):e0177678

Chapter 3

Assessment of How Users Perceive the Usage of Biometric Technology Applications

Taban Habibu, Edith Talina Luhanga and Anael Elikana Sam

Abstract

Biometrics applications are progressively widespread as a means of authenticating end-users owing to the extensive range of benefits over traditional authentication (token-base-authentication). However, the transaction involves taking into account the perceptions and responses of end-users. If end-users are fearful, hesitant about these biometric technology-applications, misuse and implementation-complications can surely overshadow. The goal of this study is to sightsee the user's-motivation, understanding, consciousness and acceptance towards utilization of biometric technology-applications. A 300-person survey was conducted to evaluate public-opinion on the use and adoption of biometrics. Stratified sample technique was used to administer the surveys. The results presented that perceived ease-of-use, user-motivation and attitude are more important-factors when deciding whether to accept new technology-applications. Although many end-users have become more familiar with biometric technology-applications (e.g., Fingerprints or facial-recognition), many individuals still have a negative-perception of the technology. Concerns regarding confidentiality and security i.e., storing and protecting personal-identification data, the fear of intruding into a person's daily-life and disclosing personal-information remain a major problem. Some end-users claim that despite the potential resilience to biometrics, designers must mentally and psychologically prepare the general public for the new use of biometric technology. This will make it possible to transform negative user-perceptions into a positive-experience. Thus, this study can help end-users and companies understand and make the right decisions to promote the use of biometric-applications and services. The study is expected to be an important research-discovery that will greatly contribute to Uganda's digital-economy.

Keywords: biometric application, user perception, privacy, security, utilization

1. Introduction

Biometric technologies are becoming more ubiquitous in our day-to-day life for a wide variety of applications such as border clearance and immigration, civilian ID cards, mobile banking, police and security, health care labs and many others [1]. The technology is used for authorization and proof of identity as a solution to the challenges associated with combatting, managing and potentially resolving criminal activity [2, 3]. In fact, mobile companies have increasingly embraced biometric technologies to allow users to connect to their mobile devices by scanning their fingerprints and faces [4]. It is estimated that 100% of mobile devices i.e., smartphones, portable devices and tablets will require biometric protection by 2020 deliberately about preventing fraud. This is quite possible because users are now exposed to biometric technologies and never realize it. Banks and credit unions have used biometrics as part of a multi-level safety means to assist address risk-related concerns. It is expected that many others will move in this direction [3, 5].

Indian, Hindustan Computing Limited (HCL) Technologies reported that e-commerce inventors are discovering the usage of biometrics and smart cards to properly prove the identity of a party to the transaction. Because it can help to reach the security facilities on the handset via voice verification [6, 7]. Since the focus is on what the user is, rather than what the user knows or possesses. The implementation of biometrics is largely dependent on the degree to which system users are willing to accept the technology [8]. User behavior may cause or break the implementation of biometric technology. The process of providing personal data publicly may be offensive to some people. As well, users may associate fingerprints with law enforcement and crime and may be unwilling to use fingerprint systems [9]. Others believe that scanning, iris and retinal systems can be harmful to their eyes. In any event, these positions may potentially contribute to significant public embarrassment to the company that collected the data, regulatory fines or law suits. If DNA scans become prevalent, they can give escalation to an entire new arena of secrecy worries such as exposure of health situations and household relationships [10].

At present, there is not a single piece of legislation that provides a comprehensive overview, addresses legislation or provides standardized guidelines for the usage of biometrics [1]. The lack of a specific document or regulation that obliges as a pre-eminent guide and governs biometric usage leaves organizations to make their own rules about how to handle and use biometric data. The potential for misuse of biometrics is an important concern for users. Consequently, the perception of the user, especially in the field of security and privacy, must be well understood. As reported by Emami et al. user's perception on use of biometric applications are generally tied to their socio-cultural, religious believe, health matters and occasionally the lawful consequence of the subject matter which has to do with delinquency. Researchers found that when applying for biometrics, individuals are unwilling to allow for instance, their faces to be captured as it violates their religious belief. Once again, others are not comfortable entering their fingerprints for fear of a security breach or for health reasons. Chandra et al. reported that while user fear is: belief, user acceptance, secrecy concerns are not taken into consideration, there is a possible threat of system failure. It might be surprising to install biometric applications without assessing the acuity of biometric knowledge [11, 12]. As a result, users must be educated on why the system was introduced and how it can be beneficial to them.

The study therefore focuses on the intensity, comprehension, awareness and acceptance of biometric use by end-users. The objective is to provide useful information and benefits of the usage of biometrics technology as well as factors affecting end-users in the usage of related technologies. The author assumes that this study will help stakeholders and policymakers at different levels to differentiate between the capacity of the application of biometrics technology, and user acceptability in the design of robust procedures for deploying biometric technologies that are user-centric. The paper is prearranged into five sections: Section 2 briefly presents several studies carried out to understand users' perceptions regarding the usage of biometric applications. Section 3 provides an idea of the method used in conducting the study. Section 4 presents the results and discusses their importance, and Section 5 presents the discussion of findings. Finally, the conclusion, limitations and some insights for future research.

2. Related work

The review addresses two main lines of research: (a) the Technology Acceptance Model (TAM) and (b) the user's perception of the usage of biometric technology application. The relevant literature for each of these two areas is discussed below.

2.1 Technology acceptance model (TAM)

To examine the public perception of the usage of biometric technology, the Technology Acceptance Model (TAM) was examined. This is an accepted model for explaining people's acceptance and behavior. Based on its simplicity and understanding [13, 14]. It helps researchers and practitioners distinguish between the reasons why a proposed technology may be acceptable or unacceptable [15]. The model is based on the Theory of Reasoned Action (TRA), a psychological approach that illustrates how the individual's belief application system acts on human behavior [16]. This implies that behavioral intent is closely related to real behavior. In essence, the TAM is based on two basic concepts: perceived usefulness (PU) and perceived ease of use (PEOU). Perceived usefulness is the extent to which a person believes that the usage of a particular technology would enhance their work performance [17]. If the assessed PU results are positive, users will tend to have confidence in the technology. However, perceived ease of use refers to the extent to which an individual believes that using a specific system would be effortless. The extent to which one believes that the usage of technology would exempt a person from conscientious work. In addition to the PU and PEOU, two other variables were expressed: attitude and motivation. Attitude is a general positive or negative assessment of a person's particular behavior. In studies of user behavior, attitude is considered as a predictor of the future inspiration to be used. Thus, the impact of the user's attitude on the intention to usage is universal, which partly explains why the TAM has been widely studied in various areas. Motivation is an indicator in which a system is used to measure subjective intent by users. This has a critical impact on whether a certain type of technology or system is accepted. Therefore, in the present study, the motivation to use was to define the magnitude of the intention of users with respect to the usage of biometric technology. **Figure 1** adds two variables that are proposed for the determinants of relative advantage, attitude and motivation to establish the intent and perception of the end-users to use the biometric technology application. The relative benefit is the level at which an innovation is better discovered than the practice previously employed. Derived from **Figure 1**, perceived usefulness, perceived ease of use, user attitudes, and user motivation are variables dependent on end-users' perception to make effective use of the application of biometric technology. As a result, technology users have greater acceptance and satisfaction. From this perspective, we anticipate the same thing in the case of accepting biometric technology. The greater the perceived usefulness, the greater the intent to accept a biometrics application system. The greater the perceived ease

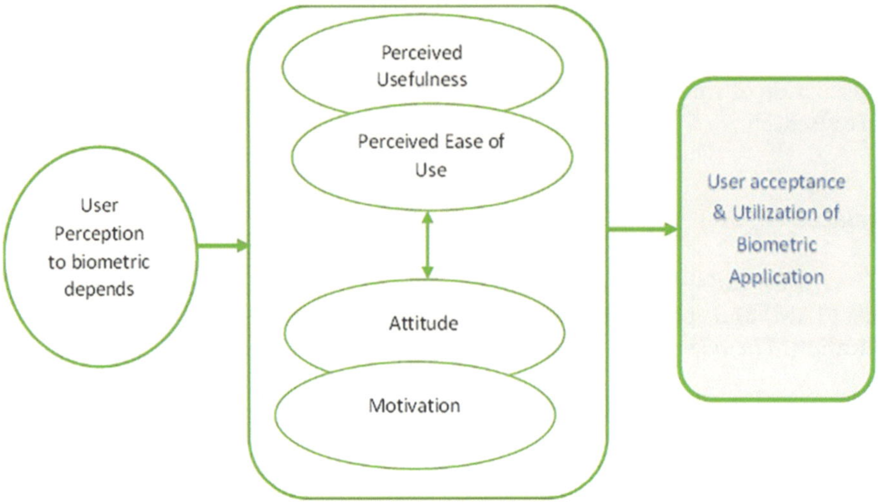

Figure 1.
Framework model for user acceptance of the biometric application.

of use, the greater the intended acceptance of a biometric system. The perceived usefulness of the biometrics system is positively correlated with perceived usability. The greater the attitude towards the use of the biometric application, the greater the likelihood that an end-user will consider a biometric application system to be useful. The higher the motivational factor, the more end users perceive a biometric application system to be easy to use.

2.2 User perception with respect to the usage of biometric applications

Increasingly, biometric technologies are being used in almost all areas of human activities for verification and identification [2, 4]. This technology allows for the collection of personal information and physiological data for identifying purposes. However, the available data is limited. Because users are more likely to have little acceptance or confidence in biometrics due to privacy concerns [18]. As such, it is significant to know the reasons that contribute to user acceptance as well as the need to consider user perception and will associated with biometric technology. As human perception is highly unpredictable in many cases, a greater comprehension of user needs is required.

Study by Habibu et al. [9] conducted a survey of user knowledge and concerns related to biometrics. The study shed light on the user's experience with the usage of biometrics. The findings present that the overall response was optimistic about their prior knowledge of biometric characteristics, but had relatively little practical experience using them. In addition, they noted that many technologies were generally better accepted than others. For example, respondents felt better about the usage of fingerprints and face images than with iris examinations. In fact, fingerprints and faces are used in many national identity systems. For example, inside access control, door pass, and client ID simply required the person to touch the sensor screen or look at the authentication device.

Carpenter et al. [19] presented a study examining workers privacy concerns associated to the organization's use of biometrics. Their findings suggest that self-determination has played a significant role in formulating privacy protection, perceived accountability, and concerns about perceived vulnerability. The research suggests that, it serve as important indicators of user attitudes to biometric technologies in the workplace.

Furthermore, a study by Jones et al. [20] explaining the purpose of users to use biometrics as an authentication tool with young Arabs were studied. The findings revealed that, perceived ease and usefulness are the most decisive factor influencing user's perception to accept or reject new technology. Therefore, the key to increasing the acceptability of any technology is to work out how the negative perceptions can be lessened.

A study presented by Chan and Elliot [21] updated biometrics secrecy perceptions with two investigations. The first investigation, carried out amongst 200 participants, asked participants of their knowledges and insights of biometrics. Another investigation, observed to measure variations in perception over time. The study suggested a level of disbelief around the safety and secrecy of the biometric data. For example, forty-five percent (45%) of participants were not able to trust their data from a public company. Because the findings revealed that there was more support for the usage of biometrics in the fight against terrorism and the banking sector.

Furthermore, El-Abed et al. [22] claimed that the major drawback in the general satisfactoriness of biometric application is the lack of general assessment method that appraises performance, users' acceptance and satisfaction, data quality and security. Such evaluation methodology assists system designer to be able to ascertain suitability of the technology being designed and aid in making necessary adjustment to the design, in the early stage, to improve the satisfactoriness level.

Study by Elliot et al. [23] reviewed technique to identify and inspect the citizen's perceptions, opinions and fears of biometrics technology. The issues such as security and privacy concerns of users are asked in the review. The findings indicated that people are pro biometrics i.e., they accept the biometrics utilization as a way to enhance security, but they have fears about their privacy (who can utilize that information). The mainstream of the individuals accepted the biometric technology, but also, have security anxieties of using biometric technology. In short, the individuals are eager to utilize the biometrics technology, but they lack hope with approximately legislative organizations. The prerequisite to teach individuals about biometrics in order to eliminate users' greatest concerns is paramount.

One common theme that comes out of the studies is that users are concerned about the privacy and security of their personal data. This is an area that requires further study as part of the proposed research, which explores the concerns of participants and the contextual nature of those concerns.

3. Materials and methods

This study involved a questionnaire survey to assess user's perception in the usage of biometric technology applications. The surveys enable to gather information to be statistically analyzed. It consisted of three sections (A, B and C). Section A was designed to capture demographic, experiential and behavioral characteristics that may affect the use of biometrics or relate to the views of participants. The participant demographic information included age, gender, the education background, the experience level about biometric technology application. The common biometric features listed in the questionnaire were fingerprint, face, iris, voice, retina, gait, signature and palm print. The analysis of the respondent's descriptive distribution is shown in **Table 1**. Section B considered questions to ascertain the participant intention, willingness and general perception with respect to the use of biometric technology applications. The five-point Likert scale from Strongly Agree (5), Agree (4), Neither (3), Disagree (2) to Strongly Disagree (1) is used. This was aimed to understand users' acceptance and utilization of biometric application. Section C considered questions to ascertain users fears in use of the biometric technology, the technique required for securing the biometric data and the strategies aimed at regulating and protecting the biometric technology information.

3.1 Analysis of the data

Data analysis involved a mixture of quantitative and qualitative techniques. Author applied Statistical method (SPSS) version 25 and presented findings using descriptive statistics in the form of frequency, percentage, mean and standard deviation to analyze responses to close-ended questions. Compared the mean independent t-test results across some aspects for instance, between user willingness and non-user willingness. A total of 300 participants (students, academic staffs and employees) from two selected institutions Muni University and IUIU University were collected. This is largely due to the fact that they are associated with a greater affinity, understanding and acceptance of new technologies, which would be necessary to transmit biometric concepts. The participants were given a consent form to notify them of

Variables	Item	Frequency	Percentage (%)
Age	21–30	94	31.3
	31–40	154	51.3
	41–50	36	12.1
	50 and above	16	5.3
Gender	Male	200	66.7
	Female	100	33.3
Education level	BSc	134	44.7
	MCs	94	31.3
	PhD	72	24.0
Role	Students	138	46.0
	Staff	98	32.7
	Employee	64	21.3
Biometric feature User experience	Fingerprint	106	35.3
	Facial	98	32.7
	Iris	30	10.0
	Retina	12	4.0
	Voice	26	8.7
	Signature	28	9.3

Table 1.
Respondents distribution frequency.

the theme and take their consent to respond in the survey. The questionnaires were provided to the participants who were comfortable in completing the survey by themselves. Stratified random sample was utilized to draw the target population. The formula $S = \frac{X^2 .\ P(1-P)}{d^2}$ was deployed for the sample size [24]. By using this approach to find the sample size, it is anticipated that the degree of bias can be fixed and the measurements of sampling error becomes low.

4. Results

4.1 Social demographic information

Out of the 300 survey participants, most of the participants were male with 66.7% and female with 33.3% respectively. The majority of participants to the survey were aged between 31 and 40 with 51.3%. Thirty-one-point-three percent (31.3%) were between 21 and 30 ages old. Twelve-percent (12%) were between 41 and 50, and 5.3% were over 51 years old. Nevertheless, this distribution of the participants' ages means that most of respondents will have either grown up with technology from an early age or been early adopters of new technologies.

In terms of education, a majority of participants had at least a high-level degree equivalent with 44.7% having at least a Bachelor's degree, 31.3% with a Master's degree, and 24% of the respondents held a doctoral degree. This is likely to be influenced by the researchers' personal and professional networks. Finally, in regards to respondent's categorical level, 46% of the participants were students, 32.7% were academic staffs, and 21.3% were employees. **Figure 2** presents the investigation of the social demographical information.

4.2 Biometrics feature utilization

The respondent's experience towards the usage of biometric technology were examined. Participants were asked about the biometric features that should be used in each of the physical and behavioral characteristics. It was used to better understand which technologies participants liked most and which ones they liked least. Participants were generally knowledgeable about a numeral of physical biometrics technologies. Thirty-five-point-three percent (35.3%) of the respondents had shared knowledge of how fingerprints are used. Thirty-two-point-seven percent (32.7%) were vast in facial scan. This is not surprising considering their commonness in personal devices and in our everyday lives (e.g., smartphones or migration at an airport). Both of these technologies have been used to protected personal devices and is increasingly become common in our daily lives. For example, the vast widely held of personal devices (e.g., smartphones and tablets) now make use of fingerprint and facial recognition so such a common usage is expected. Ten percent (10%) were having knowledge in Iris, 4% were vast in Retina.

In terms of possibly classified as behavioral biometrics technology, 8.7% were experienced in Voice, and 9.3% were vast in Signature scan. This is actually predictable in that traditionally behavioral biometrics do not require the user to interact with any specific hardware directly. Instead, their behaviors are normally monitored remotely. These analyses were pointed to the user's experience in the usage of biometric application and getting to know whether new biometric technology devices such as

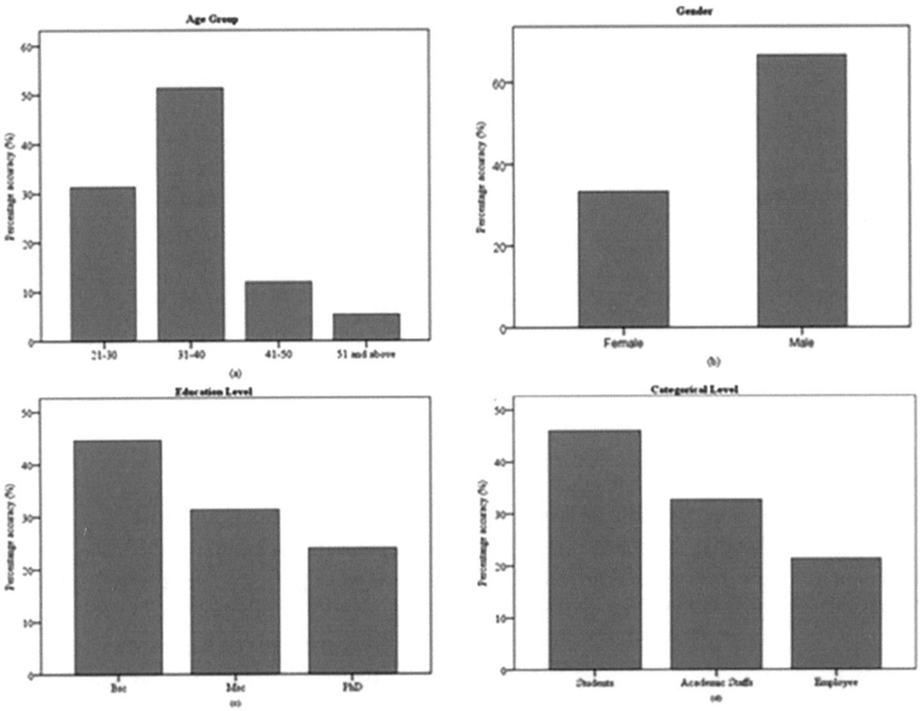

Figure 2.
Social demographic information, (a) age group, (b) gender, (c) education level, (d) categorical level.

smartphones, tablets, or laptops can either be accepted or not. Therefore, developers need to consider actual user willingness and acceptance in the utilization of these new technologies embedded with biometrics devices when designing a biometric security application, and make an effort to promote the use in a positive way. The findings from the analysis are shown in **Figure 3**.

4.3 Usefulness of biometric technology applications

Three hypotheses were verified by multiple regression analysis. Perceived ease of use (PEOU), perceived usefulness (UP), and perceived enjoyment (ENJ), with participants' attitudes towards the usage of biometric technology (ATT) as a dependent variable. Three general questions related to satisfaction with biometric technology were also raised. Sixteen quantitative questions were asked on a five-point Likert scale, ranging from strongly agree to strongly disagree. **Table 2** presents the analytical descriptive statistics for the constructions of each survey question. All three of the PEOU statements ranked highly with an average of 3.70 out of 5.00. "I would find biometric technology easy to use during workplace" scored highest with a mean of 3.73. Most of the survey statements related to perceived usefulness also ranked highly at an average of 3.66. However, respondents ranked the statement, "Biometric technology enables me to have more convenience at workplace," the lowest at 3.07. "Using biometric technology increases security level of an individual data at workplace," the highest at 4.42. With respect to

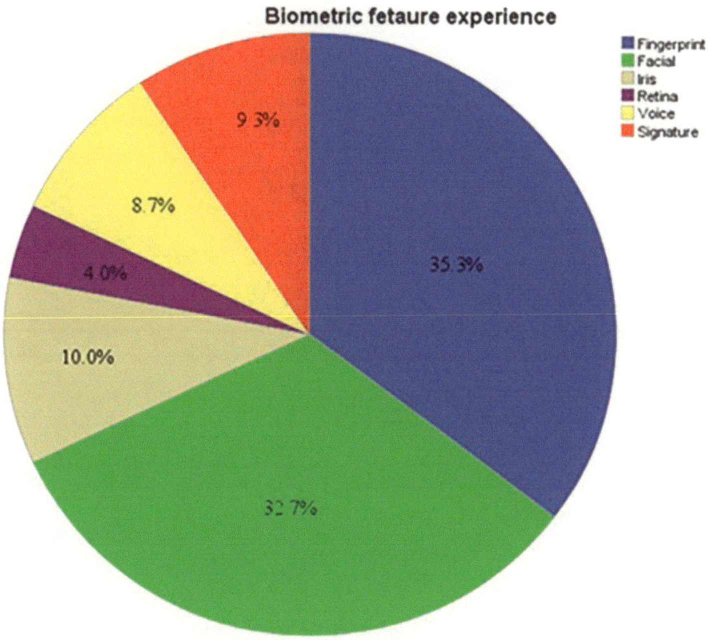

Figure 3.
Biometric feature utilization.

participants' enjoyment using the biometric technology, this category scored the lowest with an average of 2.83. "I have fun when using biometric technology" scored the lowest at 2.30 and "The actual process of using biometric technology is pleasant" scored the highest at 3.59.

Respondents' attitude towards the biometric technology was strong with an average of 3.79 out of 5.00 over the three statements. "I like the idea of using biometric technology at workplace" scored at the lowest with 3.72, while "Biometric technology makes work environment more interesting and "Using biometric technology at workplace is a good idea" were the strongest at 3.83 and 3.81 respectively. In regards to the participants' overall satisfaction with the biometric technology in general, this category scored a very high average of 3.91 over the three statements. The most highly ranked statement was, "As a whole, I am happy with the usage of biometric technology," and scored a 4.01. The results from the statistical analysis are shown in **Table 2**.

In order to gain additional insights, two open-ended questions were asked: (1) "What did you like about the biometric technology?" and (2) "What did you not like about the biometric technology?". A greater percentage 66% responded with optimistic response about the likeness of the biometric technology, while 34% of the biometric users responded with negative feedback. Of the biometric technology user group, 40% mentioned that the biometric technology was easy to use, 28% indicated greater security, while 32% showed conveniences and user friendly. Regard the negative feedback of the dis-likeness of biometric technology, 34.7% mentioned risks of personal data, 45.3% indicated that the biometric data can be stolen, while 20% mentioned insecurity of personal data.

Measurement questions	Mean	Std. Dev.	Min	Max	N	Variance
Perceived Ease of Use (PEOU) I know how to use biometric technology	3.69	1.248	1	5	300	1.551
I would find biometric technology easy to use during workplace	3.73	1.443	1	5	300	2.082
Learning to use biometric technology is easy for me	3.67	1.438	1	5	300	2.067
Perceived Usefulness (PU) I find biometric technology useful at workplace	3.18	1.278	1	5	300	1.633
Biometric technology enhances the personal security information	3.95	1.442	1	5	300	2.078
Biometric technology enables me to have more convenience at workplace	3.07	0.958	1	5	300	0.919
Using biometric technology increases security level of an individual data at work	4.42	1.043	1	5	300	1.087
Perceived Enjoyment (ENJ) I find using biometric technology is enjoyable	2.61	1.556	1	5	300	2.420
The actual process of using biometric technology is pleasant	3.59	1.369	1	5	300	1.875
I have fun when using biometric technology	2.30	1.538	1	5	300	2.365
Attitude (ATT) Using biometric technology at workplace is a good idea	3.81	1.316	1	5	300	1.731
I like the idea of using biometric technology at workplace	3.72	1.441	1	5	300	2.075
Biometric technology makes work environment more interesting	3.83	1.201	1	5	300	1.441
Overall satisfaction Overall, I am satisfied with the usage of biometric technology	3.81	1.498	1	5	300	2.243
As a whole, I am happy with the usage of biometric technology	4.01	1.168	1	5	300	1.364
I believe by attending any biometric technology conference will enhance my profounder understanding of the technology	3.91	1.430	1	5	300	2.046

Table 2.
Descriptive statistics.

4.4 User willingness vs non-user unwillingness with respect to the usage of biometric applications

In order to compare the overall user willingness vs. non-user unwillingness satisfaction levels in the usage of biometric technology, a t-test was run in SPSS. Statistical measurement of two intact groups using an independent samples t-test is appropriate to evaluate the variance amongst the two groups [24]. The results were statistically significant between user willingness vs. non-user unwillingness. The user willingness to use biometric technology mean was 4.39 and non-user unwillingness to use biometric technology mean was 3.33. This result shows that both users and non-users willingness rated their overall usage of biometric satisfaction at virtually different level. **Table 3** presents the comparison of the sample independently of the t-test results.

4.5 Security of the biometric technology

The security issues were intended to measure the extent to which subjects felt the application of biometric technology would improve the security of the end-user. Participants were asked to comment on biometric security versus other traditional methods [24]. Ninety-two percent (92%) of participants agreed with the statement that biometrics were more secure because it involves a personal presence during the verification process. Participants were also asked about the ability of biometrics to offer the same level of security as two-factor authentication. The majority 84.7% of respondents concur with this statement. Lastly, respondents were asked if they were of the opinion that biometrics could easily be compromised. Forty-eight percent (48%) of participants explained that biometrics might be compromised. While 52% stated that biometrics cannot be easily compromised, which was not surprising. This is particularly true when seeing that most respondents indicated that biometrics was as secure as two-factor authentication.

One of the key findings of this study was that participants were generally knowledgeable about the usage of fingerprints and face. This emphasizes that exposure to these technologies assists in generating support for the desired methods.

4.6 Users fear in usage of biometric technology

Another area was the level of concern of subjects about privacy issues associated with the implementation of biometric technology. The issues of willingness to provide personal biometric information for collection, use and storage were addressed. While biometric technology offers highly compelling proof of identity and individual confirmation solutions. Participants voiced concern about the usage of biometric technology, as biometrics can easily be hacked and the consequences of their mismanagement could be incredibly dangerous. Thirty-two-point-seven percent (32.7%) expressed the selling of the information to 3rd party. The danger of identity stealing is greater because, unlike a credit card, biometrics cannot be canceled or superseded if it is entered by a third party. With fingerprints all over the place and faces in full view. Forty-eight percent (48%) indicated misuse or abuse of personal data. This is because a compromised biometric data stored in the database cannot be revoked. For instance,

Comparison sample	**Mean**				
User willingness vs non-user unwillingness to use biometric technology?	Willingness(102)	Unwillingness (198)	Mean difference	t-value	Sig.
	4.39	3.33	1.059	7.610	0.000

Table 3.
Independent samples of t-test results.

the DNA information can reveal a person's health and exposure to disease [25]. Biometrics can be safely described as the future of human identification. However, this future would remain uncertain unless rigorous methods are employed to protect it from misuse or violation of data. Nineteen-point-three percent (19.3%) of the participants showed identity fraud. The analysis is shown in **Figure 4(a)**.

The manner in which personal information is increasingly collected, stored and transmitted through Internet of Things (IoT) devices and services in the cloud, make biometrics technology more susceptible to identity theft. By accessing biometric data, hackers can easily steal an individual's identity or even use and falsify personal information that may be life-threatening [26]. For example, there was an incident in 2015 when the U.S. Office of Staff Management was hacked. Cyber criminals successfully fingerprinted 5.6 million government employees and made them vulnerable to identity theft. The question of whether biometric technologies can ever be full proof remains unanswered.

4.7 Techniques for securing the biometric data

With respect to biometric security techniques, 38% of respondents recommended the use of encryption techniques as a better approach to protecting biometrics. Encryption technique can help to safeguard sensitive data. It can securely link the user's ID and biometric information to ensure that the key and biometric information cannot be retrieved from the template stored along with improving the security of personal data and communications. Increase the trust, acceptance and use of the population, which is more consistent with privacy laws.

Fourteen-point-seven percent (14.7%) expressed building centralized data Centre. This approach provides a low-cost implementation of biometric verification and benefits users who require multi-site authentication. User biometrics can be relocated over the network (generally on the Internet) and open doors for sniffing.

Twenty-four percent (24%) of the participants recommended the combination of fingerprint and strong password (Two-factor authentication). Two-factor authentication makes it possible to prevent password theft and card theft. This may make the biometrics of the database so difficult to hack or steal for a hacker. Twenty-three-point-three percent (23.3%) preferred the use of personal e-card. With smart card

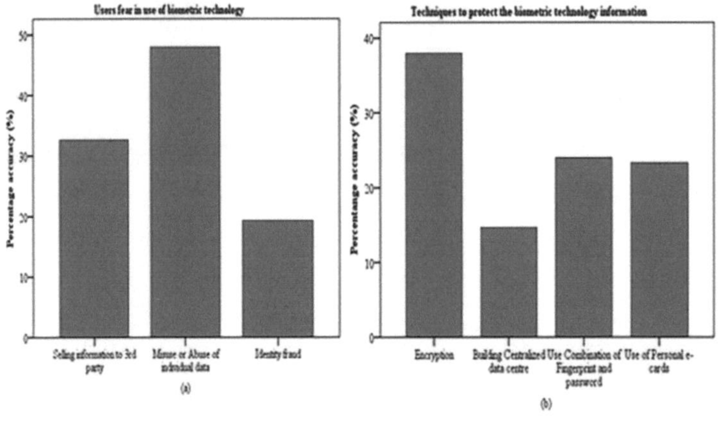

Figure 4.
(a) User fear in use of biometric technology, (b) techniques to protect the biometric data.

biometrics, users can feel that they are controlling their biometrics, increasing user acceptance of systems. In addition to these advantages, this technique also has certain deficiencies. The cost of implementing card-based biometrics is high because biometric chip readers are necessary to verify users. The user must present their biometric chip card and then their biometric credentials to the scanner to authenticate their identity. The analytical results are illustrated in **Figure 4(b)**.

In addition, other ways of securing biometrics in the database may include keeping the software up to date. When the system manufacturer informs you of an up-to-date or available software patch. It is very important to install it immediately in order to decrease the possibility of the device being susceptible to security vulnerabilities. It is particularly important to keep your operating system and Internet security software current.

4.8 Strategies to control the security of biometric information

For the aim of preventing identity theft and protecting against fraudulent acts, respondents were requested to provide a recommendation to regulate development and implementation of new security technologies as more players recognize the tremendous potential of biometric technologies. Thirty-four percent (34%) of the participants recommended that the laws to protect personal information should be adhered to. The rights of persons must be adequately protected and their data in the hands of private and public bodies must be carefully and sensitively managed. Companies needs to get consent from the individual before processing the data for surveillance or profiling [27].

Twenty percent (20%) of respondents recommended that citizens be made aware of new safety measures. Citizens need to be alert of the new security technologies put in place before being applied. The government has the ability to provide citizens with relevant information on safety and security, because it is responsible for safety and security. Through this initiative, public trust, willingness and acceptance of new biometric devices can be restored. Fourteen-point-seven percent (14.7%) of the participants recommended use of legal procedure to be taken against wrong or misuse of personal information. As biometric technologies have become more commonplace, legal procedures need to be developed to better guide employers on how to properly collect, store and use biometric information. The employer shall inform the employee if his or her biometric information is disclosed as a result of a breach of his or her records.

Thirty-one-point-three percent (31.3%) of the respondents recommended keeping the personal data as secure as possible. As the global movements show that biometric technologies are advancing rapidly, regulations need to be kept up-to-date. Technical, human, process and policy challenges need to be addressed to secure digital data and ensure that biometric technology can efficiently shape human identity authentication applications. The analytical findings are presented in **Figure 5**.

To ensure compliance with biometric security strategies, employers should:

- Audit the workplace. Perform a workplace verification to identify any biometrics used. Review the company's policies and procedures for storing, retaining, disclosing and destroying this information.

- Familiarize oneself with the law. Review applicable national and state legislation to identify legal obligations associated with the processing of employee

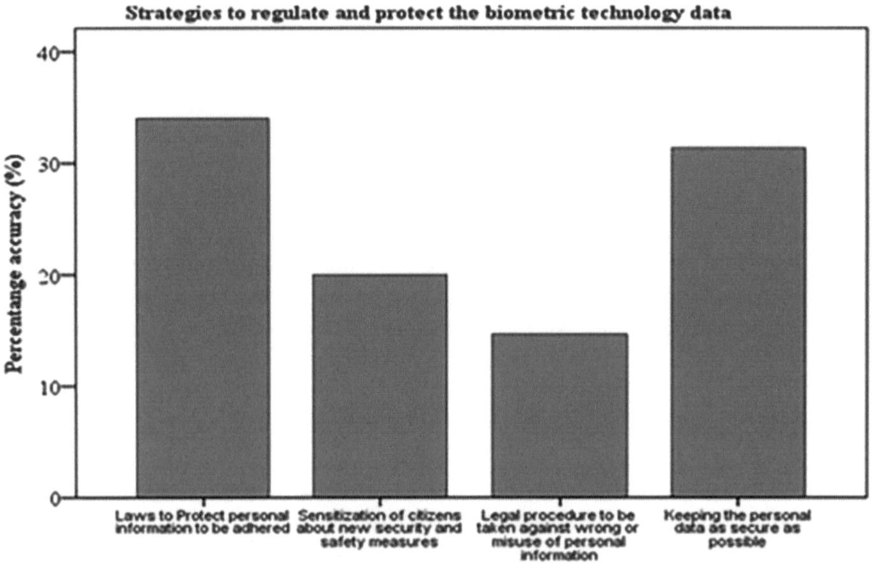

Figure 5.
Strategies to regulate and protect usage of biometric application.

biometric information. Keep in mind that obligations are expected to extend beyond employee information to include the processing of user or third-party biometric information. It's always good to consult to obtain clarification if we are not sure of the legislation that applies.

- Adopt appropriate policies. Current policies that respect state laws and inform employees of their rights and obligations as they relate to the collection, storage and use of their biometric information. One should not forget to review any data breach notification policies to ensure that they include biometric data within the definition of protected information.

- Finally, refer to service providers. A number of companies outsource certain aspects of human resources to third parties. If an employer has retained the services of a payroll company or any other supplier who collects or uses employee biometrics. It should discuss the responsibilities of the service provider with respect to this data and the efforts of the service benefactor to comply with applicable legislation. You make sure each person is on the same page when we talk about compliance.

Therefore, there is a need to guide individual awareness of the security and confidentiality of the application of biometric data collected within the daily business organization [28].

4.9 Proposed implementation of E-passport system

Two categories are defined in the biometric application process: The System Administrator task and the End-User (applicant) task. The system administrator starts by creating and deploying regional offices and agents responsible for the biometric ePassport process. With respect to the trial, thirteen (13) regional offices were deployed within the system from the four regions of Uganda. As well, the main regional headquarters controls all 13 regions. **Figure 6** shown the creation and deployment of the regions.

The system administrator assigns regional agents who are responsible for processing only the application process for that specific region. Agent is not able to see details from colleague in other regions. This is due to security concerns given that all applicants come from different regions. **Figure 7** shown how to register an agent in a given region.

The Regional Officer can only see applicants from that specific region and has no access to other applicants from other different regions or centres. The Regional Officer is responsible for approving (verifying) or disapproving (denying) the applicant as required.

Once the Regional Officer has verified (approved) the applicant, the applicant information is automatically transmitted directly to the Regional Head Office (Kampala) for further processing of the ePassport. The regional agent will no longer see the applicant information on the system. This is part of the safety measure needed for end users to track, but the officer will have a hardcopy to support the applicant information (PDF file). The applicant information is deactivated from the agent side and the Twilio SMS is sent to the applicant for status quo. In the event that the officer rejects the verified applicant information (approval) because of incomplete requirements, SMS text will automatically be sent directly to the requester for refusal to make the necessary changes and re-submit the request. **Figure 8(a)** and **(b)** shown the Twilio SMS approval and denial process for the applicant's application process.

The job of the applicant was to launch a new application by first clicking on the disclaimer acceptance button. The applicant is then guided through the on-line application process. The applicant is required to set up an account for its own credentials in the given region. Here, an applicant will be provided with a randomly generated Unique Identification Code (UIC) from the selected region. The information concerning the applicant's credentials are shown in **Figure 9**.

The applicant will then proceed to fill in his/her personal details, the origin, residential, spouse detail if any, the person of contact (next of kin), and finally the parents' details. Once the information is provided, the applicant will provide the required attachments, such as a personal passport photograph. This passport photograph will be automatically encrypted for security reasons. This will only be visible to the requester if the correct password is entered. Keep in mind that you will not be able to see your passport picture when you are not using the system. Other documentation included a copy of the National Identification Number (NIN), a letter of reference

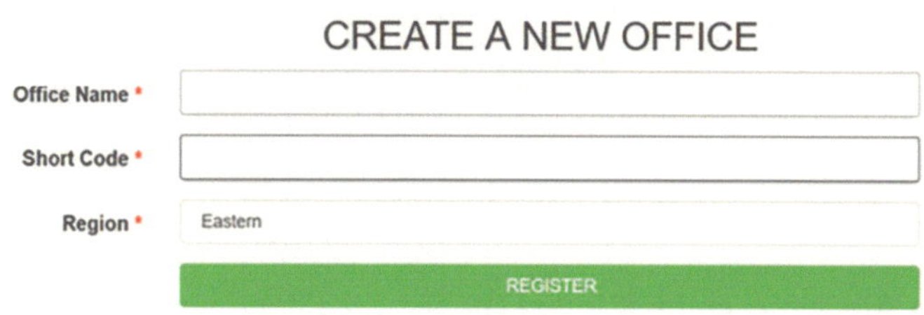

Figure 6.
Creation and deployment of regions.

Figure 7.
Register an agent in a given region.

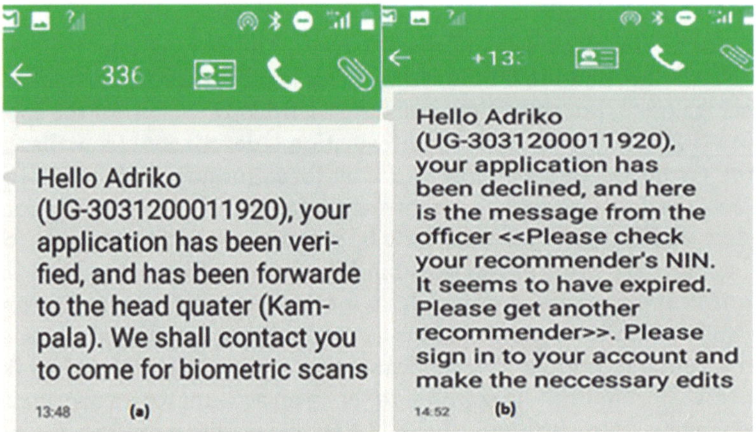

Figure 8.
(a) Verified applicant information, (b) denial of applicant information.

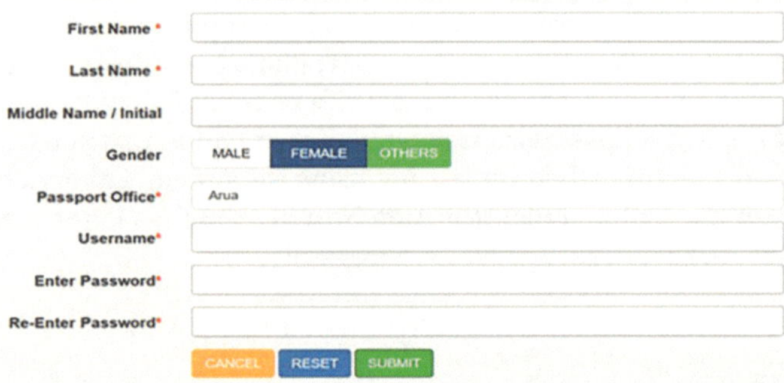

Figure 9.
Applicants login details.

and a payment slip. Upon completion of the required documents, the applicant can now submit the form. Prior to the submission of the application, the applicant can download a copy of the application details for further investigation or follow-up and backup.

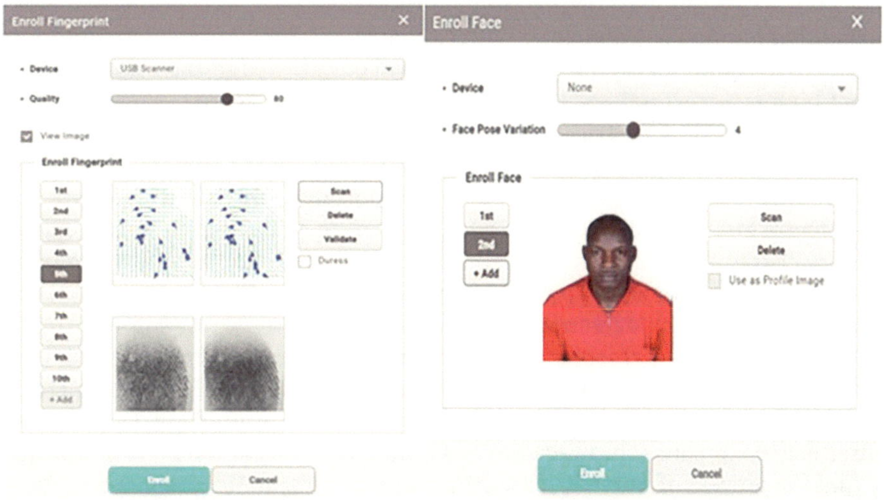

Figure 10.
Biometric fingerprint and ficial enrollment.

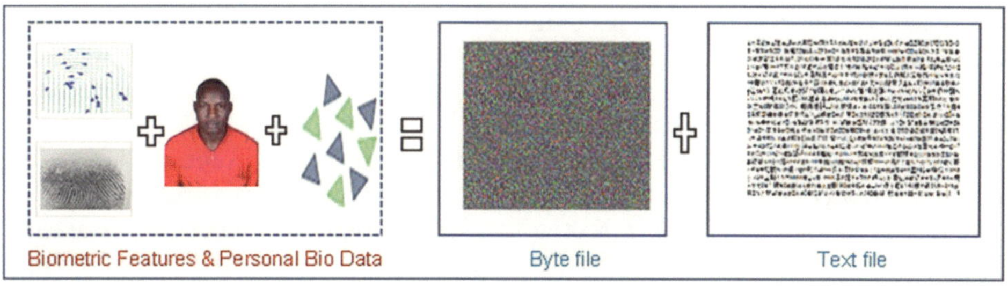

Figure 11.
Encoded byte and text files.

The features of biometric technology used were fingerprints and a face image. The authors used the Biostar 2 Server standard to retrieve biometric features and templates. **Figure 10** shown the enrolment in biometrics.

The enrolled biometric function and the personal information were combined and encrypted to produce two files, i.e. the byte file and the text file for the security reasons. This ensures that no trespasser may have access to personal information without his or her consent. It was practical evidence for the end user to understand and motivate themselves on how users perceive the use of biometric technology applications. Everyone has the privilege of reviewing the status of their claim. Any unlawful attempt with individual data, a text message is automatically sent to the individual for warning, detection and alert. The organization is in a position to handle individual data and provide timely feedback. Consequently, the increased process of biometric passport system, accuracy and competence of users. **Figure 11** shown the encrypted text and byte files.

5. Discussion of the findings

The study focused on user perception when using biometric technology to characterize its operationality and acceptability. It determines whether a meaningful relationship exists amongst demographic features and user perceptions. Participants revealed that biometric technology is more secure and appropriate than traditional approaches to fraud. It also revealed a good level of knowledge and acknowledgement of biometric technology. Participants had first-hand acquaintance of a numeral of physical biometrics technologies. It ought to be noted that while these technologies are widely accepted, they depend heavily on the context. However, the study revealed that users appear to be most comfortable with these biometric feature (e.g., fingerprints and face recognition). Considering somewhat more intangible biometric characteristics (e.g., voice and signature) they tend to be less popular. This proposes that there might be opportunities to increase public acceptance of these behavioral biometric characteristics.

User willingness and acceptance of biometric technologies were reviewed. We conclude that acceptance of biometric technology might be highly dependent upon the degree to which system users are willing to accept the technology. The attitude of the user may do or pause the operation of a biometric application. Some individuals may find the process of publicly disclosing personal information unpleasant. As well, users may associate fingerprints with law enforcement and crime. Others may think that iris or retinal systems can harm their eyes, despite clear evidence to the contrary. Therefore, it is necessary to educate users about why the system was introduced and how it could be beneficial to them [29]. The study also examined the age of participants based on their opinions on the usage of biometric technologies. The majority of respondents were aged between 31 and 40 with 51.3%. This distribution means that most of respondents will have either grown up with technology from an early age or been early adopters of new technologies.

Overall, the findings of the study indicated that perceived usefulness (PU) was the highest forecaster of user attitudes to the usage of biometric technology. This outcome is consistent with a number of other studies that demonstrate PU as one of the strongest predictors of attitude and confirms importance to explain the users' attitude towards event biometric technology.

However, the study has some limitations concerning the scope of the study. The survey participants consisted of two institutions within a limited geographical province of the northern part of Uganda, limiting the regional diversity. A future study could extend the spatial reach by surveying users' perception to the usage of biometric technology applications across all regions of the Uganda or East African Community (EAC) member states or even globally. Furthermore, the vast majority of the study's participants were male (66.7%). While still offering valuable insights, the study could have benefitted from a better male/female balance. Another limitation of this investigation is the data collection method. While the responses were collected from both students, staffs and employees of the institution, this study did not compare the demographic profile of each group before combining data. In reality, staffs and students of the institution might have different characteristics. Thus, it would be worthwhile if future research compares staffs and students' profiles or focuses on one population. Furthermore, the study neglected to ask the non-user unwillingness to use biometric technology any open-ended questions. Future research could delve into the non-user unwillingness for rejecting the usage of biometric technology application by asking an open-ended "why?" questions. This study only included three variables (e.g., perceived usefulness, perceived ease of use, and perceived enjoyment) as pre-cursors of attitude. In addition, the finding indicated that attitude is a significant predictor of satisfaction in the usage of biometric application. Thus, future studies could incorporate more variables that might contribute to the variance in attitude and satisfaction (e.g., prior experience) into the current model. Future work in this area would further develop a deeper understanding of the experiences and perceptions of biometrics technologies amongst the general public. This could be achieved by larger more representative samples, and cover wider ranges of biometrics and related technologies which we were unable to address in this article. There are other limitations which could also form the basis for future work in this area to transcend the work conducted here. Notably, the conceptual framework should be tested on other samples and on samples as representative as possible of the whole population in order to see if all hypotheses postulated can be verified. In addition, it would also be appealing to examine the motivations of public and private organizations, their perceived risks in the implementation of biometrics, and the effects of private companies or public bodies in enabling this process (e.g., e-government initiatives such as e-passport).

The survey results, also indicated that same applicant may be reluctant to move to the biometric application system. However, they seemed eager to admit the application technology when requiring the biometric permit from the migration office. Those who have accepted biometric application are pleased with the relative ease of use, convenience, and increased security offered. Raising awareness and overcoming goals such as privacy and data safety issues will help promote wider acceptance of the technology. As their comfort level with using the biometric application in the regional Centre's increases, they may be more eager to welcome the system when processing the permit. Many users probably do recognize the value of the biometric application to them. Study results indicated that users were not totally convinced that biometric application could speed up the process of the permit and congestion in the waiting stages. User education is probably critical in accelerating individual acceptance of the application. This survey has revealed some strategies (i.e., Quick application, free status checks, notification SMS for security purpose). However, extra investigation is required to determine strategies that will cause users to use the application on a frequent basis.

The study suggests that the greatest benefit to users who use the application systems will be increased privacy and security, easy permit process time and SMS notification for violation of the personal data. Users will not face the danger of having their data be compromised or stolen and used by someone else. Since any attempt by the user's information, will notify the user immediately and he/she can report the incidence to police to make a follow up. However, this benefit evaporates if only users use strong authentication character during the process of creating their user accounts. Users would still have to remember their credential for checking the status of their permit. It will probably take a wider implementation by many users to insure this promised benefit for all. The study further, proposed encoding methods as the most favorable tactic of protecting the biometric data application, because the encryption produced two encoded files (byte and text file) to securely protect data where no one else has access. These files are incorporated with Twilio message. The message is auto-generated directly from the application database, to alert users in circumstances where an attacker tries to access the application database. The text message is one of the security mechanisms successfully implemented. It helps inform the users and the authority, how secure and safe the individual biometric data application.

Since there is little research conducted concerning biometric passport application for end-users, the area provides a means of opportunities for further research. For example, can users identify who the intruder when data is compromised? Can they have privilege to deny access when the data are used for another purpose other than to purpose required for? Can the same application be utilized to detect when employees arrive or leave? There will likely be many possible uses, and users will need research data that will help make the decision on or not to implement the application. Researchers need to determine how much privacy users are eager to bounce up for faster service and more security. Until these privacy footraces are overcome, biometrics application may have a hard time getting a position in most database storage. Further, significantly, do users perceive greater benefits? Users need to be persuaded that the application knowledges offer extra rewards than current application systems. The study, therefore, recommended that, greatest practices are required for the strategy and growth of biometric application and the procedures for their action. Secondly, social, legal, and cultural factors can affect the acceptance and effectiveness of biometric application systems and must be taken into justification in application design, development, and deployment. Ideas of proof related to biometric application authentication must be built on solid, peer-reviewed trainings of system accuracy under many conditions and for many persons reflecting real-world sources of error and uncertainty in those mechanisms. Companies will need to assure potential users that their data is safe, using methods such as high-level encryption and hardened data Centre's, and aggressively promote these features. An increase in testing and input from end-users in the design phase may also help to raise awareness and make the transition from knowledge-based to biometric security easier. This is particularly significant in how it relates to making the application easy to use, learn, and rely on. There should be a need for law-making to guard in contrast to the robbery or deceitful usage of biometric application systems and biometric data. Since governments are both major producers and utilizers of citizen identity data, as public authorities, they have a responsibility to guard the secrecy of those they represent. They use biometrics to provide efficient and secure access to citizen services, through reliable identification of individuals. The public sector is currently the main marketplace for biometric applications worldwide, such as identity cards, social benefits, immigration and border control, or e-voting. They are typically bound by the same privacy laws as private sector organizations. If anything, they face a greater degree of scrutiny from regulatory bodies such as Data Privacy Commissions, not only for the vast amount of individual data to which they have access, but also to fears of a "surveillance state".

6. Conclusions

In this paper, user's embracing and gratification of biometric application deployment is explained. The findings indicated that users have the willingness and high acceptability in using the biometric application, but they have worries relating to the biometric data infringement, such as selling of individual data to 3rd parties, misuse or abuse of individual data and identity fraud. The proposed encryption algorithm helped build users' confidence. The encryption encoded user's biometric data kept in the database and other roles in the application. User biometric data application in the storage is highly secured and protected with Twilio SMS. And prevented individual data from been compromised by an impostor, hence higher security of individual privacy data. The application will help users to understand the process of applying online. It has capabilities of conducting fraudster inquiry based on personal data, retrieval and dissemination of security information. This study will help designers to solve security-usability-privacy trade-offs when designing biometric technology application features.

Acknowledgements

The authors would like to acknowledge Nelson Mandela African Institution of Science and Technology, Arusha-Tanzania. Muni University Arua-Uganda, and Hamitech Computer Centre Limited for the financial support and research resources.

Conflicts of interest

The authors declare no conflict of interest.

Author details

Taban Habibu[1*], Edith Talina Luhanga[2] and Anael Elikana Sam[2]

1 Department of Computer Science and Electrical Engineering (CSEE), Muni University, Arua, Uganda

2 School of Computational and Communication Sciences and Engineering (CoCSE), Nelson Mandela African Institution of Science and Technology (NM-AIST), Arusha, Tanzania

*Address all correspondence to: hamitech2019@gmail.com

References

[1] Biometrics and Fraud: You're One in 7 Billion. 2019. Available from: http://www.paymentscardsandmobile.com/biometrics-and-fraud-youre-one-in-7-billion [Accessed: May 01, 2020]

[2] BBC. Banks Turning to Voice Recognition. 2016. Available from: http://www.bbc.co.uk/news/business-36939709 [Accessed: February 19, 2019]

[3] Bank Customers to Use Biometrics by 2021: Goode Intelligence. 2018. Available from: https://findbiometrics.com/bank-customers-biometrics-2021-goode-intelligence-509244 [Accessed: May 02, 2020]

[4] Buckley O, Nurse JRC. The language of biometrics: Analysing public perceptions. Journal of Information Security and Applications. 2019;**47**:112-119. DOI: 10.1016/j.jisa.2019.05.001

[5] Namiti A, Ondiek DCO. Adoption of biometric system to manage teachers absenteeism for improvement of teachers performance: A case study for Karuri High School in Kiambu County, Kenya. International Journal of Scientific and Research Publications. 2020;**10**(05):434-445. DOI: 10.29322/ijsrp.10.05.2020.p10150

[6] Mwapasa M et al. 'Are we getting the biometric bioethics right?'—The use of biometrics within the healthcare system in Malawi. Global Bioethics. 2020;**31**(1):67-80. DOI: 10.1080/11287462.2020.1773063

[7] Singh B, Singh N. A mobile app to make biometrics usage more secure. Tech Powered by IEEE. 2020;**5**(42):1-29

[8] El-Abed M, Giot R, Hemery B, Rosenberger C. A study of users' acceptance and satisfaction of biometric systems. In: 44th Annual 2010 IEEE International Carnahan Conference on Security Technology. French: HAL Open science; Vol. 2010. pp. 170-178

[9] Habibu T, Luhanga ET, Sam AE. Evaluation of users' knowledge and concerns of biometric passport systems. Data. 2019;**4**(April):1-17. DOI: 10.3390/data4020058

[10] Zirjawi N. A survey about user requirements for biometric authentication on smartphones. IEEE. 2015;**2**(12):1-6

[11] Cornacchia M, Papa F, Sapio B. User acceptance of voice biometrics in managing the physical access to a secure area of an international airport. Technology Analysis & Strategic Management. 2020;**32**(10);1-15. DOI: 10.1080/09537325.2020.1758655

[12] Chhabra V, Rajan P, Chopra S. User acceptance of new technology in mandatory adoption scenario for food distribution in India. International Journal on Food System Dynamics. 2020;**11**(2):153-170. DOI: 10.18461/ijfsd.v11i2.47

[13] Habibu T, Luhanga ET, Sam AE. Developing an algorithm for securing the biometric data template in the database. International Journal of Advanced Computer Science and Applications. 2019;**10**(10):361-371

[14] Emami C, Brown DR, Smith DRG. Use and acceptance of biometric technologies among victims of identity crime and misuse in Australia. Trends and Issues in Crime and Criminal Justice. 2016;**10**(511):1-6

[15] Chandra A, Calderon T. Challenges and constraints to the diffusion of biometrics in information systems. Communications of the ACM. 2005;**48**(12):101-106

[16] Jin CC, Seong LC, Khin AA. Factors affecting the consumer acceptance towards Fintech products and Services in Malaysia. International Journal of Asian Social Science. 2019;**9**(1):59-65. DOI: 10.18488/journal.1.2019.91.59.65

[17] Dhagarra D, Goswami M, Kumar G. Impact of trust and privacy concerns on technology acceptance in healthcare: An Indian perspective. International Journal of Medical Informatics. 2020;**141**(February):104164. DOI: 10.1016/j.ijmedinf.2020.104164

[18] Miltgen CL, Popovič A, Oliveira T. Determinants of end-user acceptance of biometrics: Integrating the 'Big 3′ of technology acceptance with privacy context. Decision Support Systems. 2013;**56**:103-114

[19] Pai CK, Wang TW, Chen SH, Cai KY. Empirical study on Chinese tourists' perceived trust and intention to use biometric technology. Asia Pacific Journal of Tourism Research. 2018;**23**(9):880-895. DOI: 10.1080/10941665.2018.1499544

[20] Thongsri N, Shen L, Bao Y. Investigating academic major differences in perception of computer self-efficacy and intention toward e-learning adoption in China. Innovations in Education and Teaching International. 2019;**00**(00):1-13. DOI: 10.1080/14703297.2019.1585904

[21] Chien S-Y, Lewis M, Hergeth S, Semnani-Azad Z, Sycara K. Cross-country validation of a cultural scale in measuring trust in automation. Proceedings of the Human Factors and Ergonomics Society Annual Meeting. 2015;**59**(1):686-690

[22] Carpenter D, McLeod A, Hicks C, Maasberg M. Privacy and biometrics: An empirical examination of employee concerns. Information Systems Frontiers. 2018;**20**(1):91-110

[23] Jones LA, Antón AI, Earp JB. Towards understanding user perceptions of authentication technologies. WPES '07: Proceedings of the 2007 ACM Workshop on Privacy in Electronic Society. 2007;**13**(6):91-98. DOI: 10.1145/1314333.1314352

[24] Shalabh K. Chapter 4 stratified sampling. In: Stratif. Sampl. Helvetic Editions LTD. Switzerland: International Journal of Academic Research in Management (IJARM); 2014. pp. 1-27. Available from: http://home.iitk.ac.in/~shalab/sampling/chapter4-sampling-stratified-sampling.pdf

[25] Elliott SJ, Massie SA, Sutton MJ. The perception of biometric technology: A survey. In: 2007 IEEE Workshop on Automatic Identification Advanced Technologies. Alghero, Italy: IEEE; Vol. 2007. pp. 259-264

[26] El-Abed M, Giot R, Hemery B, Rosenberger C. A study of users' acceptance and satisfaction of biometric systems. Proceedings—International Carnahan Conference on Security Technology. 2010;**36**(99):469-476. DOI: 10.1109/CCST.2010.5678678

[27] Wilson C et al. Fingerprint vendor technology evaluation 2003: Summary of results and analysis report. NIST Interagency/Internal Report (NISTIR). 2004;**7123**:1-79

[28] Habibu T, Luhanga ET, Sam AE. A study of users' compliance and satisfied utilization of biometric application system. Information Security Journal. 2020;**30**(3):125-138. DOI: 10.1080/19393555.2020.1813354

[29] Bustard J. The impact of EU privacy legislation on biometric system deployment: Protecting citizens but constraining applications. IEEE Signal Processing Magazine. 2015;**32**(5):101-108. DOI: 10.1109/MSP.2015.2426682

Chapter 4

Behavioral Biometrics: Past, Present and Future

Mridula Sharma and Haytham Elmiligi

Abstract

Behavioral biometrics are changing the way users are authenticated to access resources by adding an extra layer of security seamlessly. Behavioral biometric authentication identifies users based on a set of unique behaviors that can be observed when users perform daily activities or interact with smart devices. There are different types of behavioral biometrics that can be used to create unique profiles of users. For example, skill-based behavioral biometrics are common biometrics that is based on the instinctive, unique and stable muscle actions taken by the user. Other types include style-based behavioral biometrics, knowledge-based behavioral biometrics, strategy-based behavioral biometrics, etc. Behavioral biometrics can also be classified based on their use model. Behavioral biometrics can be used for one-time authentication or continuous authentication. One-time authentication occurs only once when a user requests access to a resource. Continuous authentication is a method of confirming the user's identity in real-time while they are using the service. This chapter discusses the different types of behavioral biometrics and explores the various classifications of behavioral biometrics-based on their use models. The chapter highlights the most trending research directions in behavioral biometrics authentication and presents examples of current commercial solutions that are based on behavioral biometrics.

Keywords: behavioral biometrics, gait, mouse dynamics, keystroke dynamics

1. Introduction

Multi-factor authentication is a promising authentication method, in which the user is required to provide two or more verification factors to gain access to a service or a resource. Multi-factor authentication could use One time Passwords (OTPs), physical biometrics such as face-recognition or finger-prints, etc. Although passwords have been used regularly for authenticating users for years, they are losing their popularity as passwords can be cracked or stolen quite easily. Biometric security was introduced as a better solution to verify individuals based on their unique characteristics [1]. Physical biometrics, such as fingerprints, face recognition and iris scanning, are currently being used extensively in many applications to secure access to servers and services. However, they are mainly used to perform static authentication to grant access to authorized individuals. Physical biometrics are not commonly used to constantly authenticate users while they are using the service.

With the escalating cybercrimes, static authorization fails to keep systems secure. Session hijacking and man-in-the-middle attacks are just two examples of possible threats that can have significant impacts on systems and networks, even if static authentication was deployed. Therefore, security experts are currently considering the implementation of dynamic, continuous authentication in a wide range of applications. Continuous authentication can be done using behavioral biometrics (BB), which is one of the most promising solutions to this problem. Also known as *behaviometrics*, it is the future of user authentication as it provides a secure, seamless, and hassle-free digital experience. Behavioral biometric authentication systems are currently being deployed in banks, government organizations, and other facilities to provide an efficient protection system against cybercrimes [2].

Since behavioral biometrics is a continuous way of authentication, it keeps checking the behavioral patterns of users. Body movements, voice modulations, typing style and speed, mouse movement styles, and behavior are some of the behavioral biometrics which are known to have uniqueness in it. The behavioral biometrics are primarily based on either the way human-computer interactions take place or the measurements of the body parts and muscle actions [2]. It focuses on how a user conducts a specific activity rather than focusing on an activity's outcome [3].

1.1 Chapter road-map

This chapter begins with an overview of behavioral biometrics in Section 2, which discusses the different types of behavioral biometrics, their advantages, and their shortcomings. Section 3 provides a survey of the research work on behavioral biometrics in the literature. This includes the latest research trends and directions related to behavioral biometrics. There are also several industrial organizations providing commercial platforms that support behavioral biometrics authentication. Section 4 provides a review of those companies and their products. Section 5 presents cases studies of various application domains where behavioral biometrics is used for security authentication. Finally, we draw our conclusion in Section 6.

2. Behavioral biometrics: what and why?

With the increasing level of fraud and unauthorized intrusions in various areas of life, especially in banking; the need of multi-factor authentication was significant. Companies and service providers started enforcing multi-factor authentication as a new security requirements to maintain access to services or resources. Biometrics are currently used in many applications as the second level of authentication, along with passwords, for authorizing or even identifying users.

2.1 Behavioral biometrics vs. physical biometrics

There are two main categories of biometrics that are currently being used. These two categories are physical and behavioral. Physical (physiological) biometrics depends on the measurements of a specific individual's features for identity verification/authentication. This includes face geometry, fingerprints, certain parts of the eye, vein patterns, and other corporal traits. To put it simply, physical biometrics replace "things that you know" (passwords and PINs) with "things that you are" [4]. Other examples include DNA, ear, footprint, palm print, retinal, etc.

On the other hand, behavioral biometrics is the measurement and analysis of human-specific behavioral traits based on human movement or their interaction with the computer parts, such as mouse, keyboard or handheld devices like ipads, or phones.

Physical biometrics are commonly used for one-time authentication, whereas, for dynamic authentication, behavioral biometrics can be more effective. Behavioral biometrics deployment can be divided into four distinct types of applications: continuous authentication, risk-based authentication, insider threat detection, and fraud detection and prevention [3, 5]. Behavioral biometric authorization integrates three main fields: human behavioral pattern analysis, smart sensors technologies, and machine learning models.

The biometric types are shown in **Figure 1**.

2.2 Advantages of behavioral biometrics

There are many advantages of behavioral biometrics over physical biometrics. The following points highlight these advantages [5–7].

- ***Continuous collection and authorization***—Behavioral biometrics enable constant monitoring of users. This helps to ensure that only the authorized user is the one who is using the system, even after the initial identity check has been done.

- ***Non-obtrusive collection***—The behavioral data can be collected in a seamless manner without disturbing the normal service usage.

- ***No need of special hardware***—The behavioral data may be collected using a standard camera or voice recorders. The video or audio recordings are processed to retrieve the data for authorization afterward.

- ***Useful for authorization***—Behavioral biometrics deliver continual user authentication and is a powerful defense. But it is only a complement to one-time

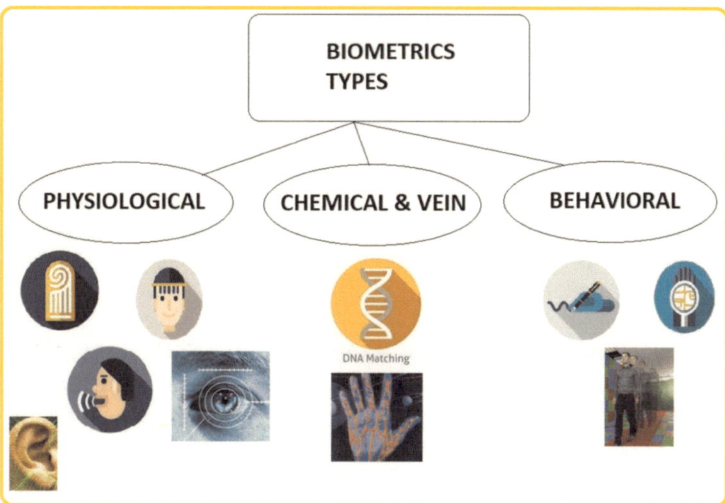

Figure 1.
Types of biometric.

authentication techniques such as passwords, PIN, and other physiological biometrics.

- ***Universality***—When applied to a large population, the universality of behavioral biometrics is very low as the degree of difference in behaviors may not be very large. But when used in a specific domain, the actual universality of behavioral biometrics reaches up to 100%, making it highly acceptable.

- ***Circumvention***—Behavioral biometrics traits are very difficult to emulate or copy.

- ***Unique combination***—Behavioral biometrics is mostly a unique combination of analyzed behavioral characteristics for each real person.

- ***Smooth Integration***—Once the behavioral biometrics model is defined, it can be integrated very easily with already existing security systems. For example, the regular video surveillance system can be utilized to implement behavioral biometrics system.

- ***High verification accuracy***—In multi-modal identification systems, the behavioral biometrics verification accuracy is proven to be quite high.

- ***Acceptability***—Most often, behavioral biometrics are collected without user participation. Therefore, it does have a high degree of acceptability. However, on privacy and ethical grounds, it faces several objections as well.

2.3 Shortcomings of behavioral biometrics

Although behavioral biometrics authentication has high accuracy and acceptance rate, it still has several challenges that hinder the implementation of such systems in a wide range of applications. The following points highlight these challenges.

- ***Implementation Cost***—Although, the new hardware is not required, still a framework that can create the dataset for behavioral biometric analysis needs to be built and integrated separately into the existing security systems. The implementation of such a new framework can be costly since it is still in the development stages.

- ***Large Data Acquisition***—The integration of behavioral biometrics authentication requires the collection of huge personal data records to profile a user's typical behavior accurately.

- ***Adaptation to Behavioral changes***—One of the biggest challenges is the ability to create a classification model that can adapt to behavioral changes. Changes in human behavior can happen for many reasons, such as external factors like weather, tiredness, or even aging. Behavioral biometrics authentication models need to be constantly re-trained to be up to date with the changes in human behavior. People may behave differently when they are in a hurry, tired, drunk or when they are not feeling well. Behavioral biometrics models face many challenges related to the adaptation to behavioral changes.

- ***Privacy Issues***—Some users are still reluctant to use behavioral biometrics authentication due to ethical and privacy issues.

2.4 Commonly used behavioral biometrics

Behavioral biometrics systems measure various human actions. These actions can be the result of human skills, such as motor skills, style, preference, knowledge, or strategy [5]. Based on the traits and features used for collecting human behavior, behavioral biometrics can be classified as:

- ***Skill-based Behavioral Biometrics***—The behavior is based on the instinctive, unique and stable muscle actions taken by the user. Examples are car driving style, keyboard dynamics, programming style, gaming, etc.

- ***Knowledge-based Behavioral Biometrics***—The knowledgeability of the user is recorded as their usual behavior. Examples are biometric sketch, text authorship, etc.

- ***Style-based Behavioral Biometrics***—Each user has a unique style that can be used to authorize them. Examples are haptic, gaming, programming, mouse, painting, email behavior, gesture etc.

- ***Strategy-based Behavioral Biometrics***—Users may have a specific strategy that they adopt. An example is the gaming technique.

- ***Preference-based Behavioral Biometrics***—Based on the user's preference of words, letters, or their belongings. Examples are credit card usage, bank usage, tool usage, language usage etc.

- ***Motor-skill-based Behavioral Biometrics***—Based on the muscle-control actions of the users makes it innate, unique, and stable. Examples are blinking, GAIT, handgrip, haptic, lip movement, signature, tapping, voice/speech, etc.

2.5 How does behavioral biometric authentication work?

For the purpose of identification or authorization, behavioral biometrics data is first collected and stored. The data is processed further to prepare a signature profile. Using machine learning classifiers, predictive models are trained, developed, and evaluated. Later, this model is used as a comparison tool, whenever the user uses the application. Using behavioral patterns, the model is used to continuously verify the user's profile throughout their working sessions. The generic architecture of a biometric system consists of five main modules:

- Data Collection Module: This module captures the biometric raw data to extract a numerical representation.

- Feature Engineering Module: To reduce the extracted numerical representation and optimize the data into required features that need to be stored for the verification and identification purposes.

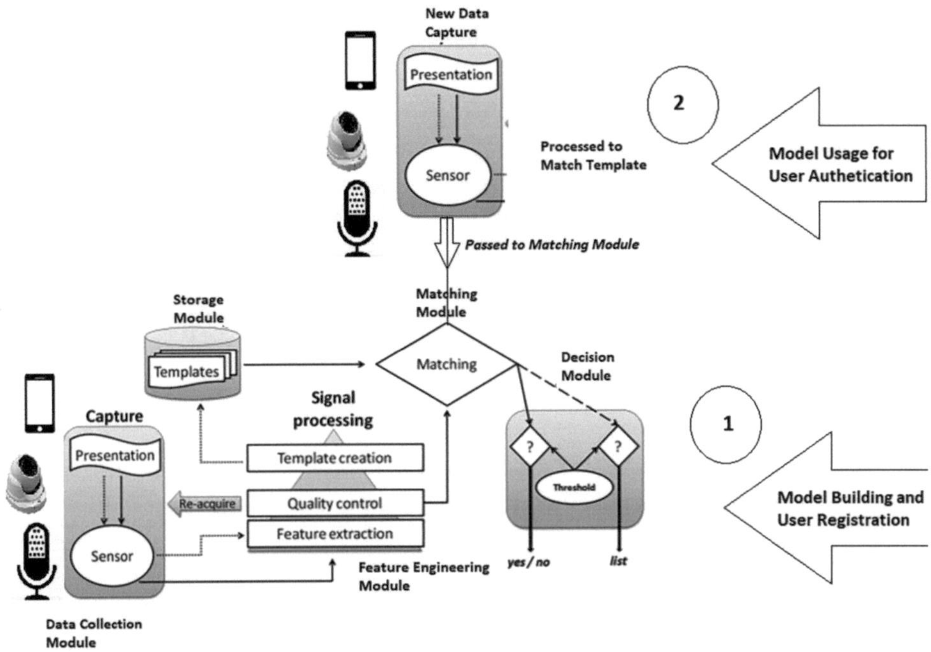

Figure 2.
Behavioral biometric model.

- Storage module: This module stores the individuals' biometric profiles in the form of dataset.

- Matching module: The module is used to compare the newly extracted biometric profile to one or more previously stored profiles.

- Decision module: This is the verification step to return a value that decides for identification/authorization.

The BB model is shown in **Figure 2**.

3. Behavioral biometrics models in literature

Behavioral biometrics has drawn the attention of both researchers and industry experts. The common areas where behavioral biometrics has played a very important role are user profiling, user modeling, opponent modeling, criminal profiling, jury profiling, etc. [5]. The information/data that may be collected for behavioral analysis may come from several sources like sensors, cameras, keyboard and mouse usage, device, audit logs, signatures or handwriting, programming style, language, smell, etc. [5]. Moreover, physical traits like odor, heartbeat, and even DNA are also being used in some applications. Researchers have also started exploring ECG, brainwaves, and passthoughts to analyze behavioral traits [5].

The most commonly used behavioral biometrics is keystroke dynamics. Keystroke dynamics have been used to authenticate users for years. Keystroke dynamics data can be collected by typing standard or non-standard passwords. Features extracted from the raw data that represent the typing patterns are used to create a unique profile for each user and to authorize those users later to resources [8–10]. It can also be used to recognize the emotions of a person [11]. To recognize the emotion from typing patterns, users are asked to type a specific sentence. Using feature extraction techniques, predictive models can be trained to classify various emotions. In one study, touch sense was defined and created as an emotion detection model based on typing and swiping patterns of a user with an accuracy rate of 73% [12]. Typing and swiping patterns are used in several applications to detect the emotions of smartphone users [12].

Another example of behavioral biometrics is mouse dynamics, where the recognition of a user profile is done based on the way a user uses his/her mouse on the computer [13–15]. The behavioral profile is created by extracting specific features related to the mouse movements of a user. Mouse and keystroke dynamics are related and complement to each other. The use of the mouse is very important in graphical user interface applications, while the keyboard is commonly used in word processing and command-line applications [16]. Mouse and keystroke dynamics are significantly important in enhancing computer security.

One of the most interesting research directions in behavioral biometrics is GAIT analysis. GAIT analysis is used to authenticate users based on their style or manner of walking [17, 18]. GAIT analysis systems depend mainly on a video camera, that captures images of people walking within its field of view. The images are processed to get appropriate features of users such as joint angles or silhouettes and the values are then compared to the stored gait signatures and profiles of the authorized individuals. One of the main advantages of GAIT analysis is that it is non-intrusive, which means that it does not require cooperation from the individual, and can function at moderate distances from the individual under observation.

Biotouch is another framework based on behavioral biometrics and location for continuous authentication on mobile banking applications [19]. Biotouch uses touch patterns for profiling users while typing and holding the device. This data is then used for predictive model building and authorization.

A new technique in profiling users' behavior is creating users' profiles based on their game playing styles. This technique analyzes the strategies used while playing a game and creates a user profile based on these strategies, as a type of behavioral biometric. These profiles are used later for continuously observing and authorizing the player to the servers [20]. One example of using this new technique is exploring the strategies used while playing the poker game to create behavioral biometric profiles [20]. Once a profile is created, it can be used to authorize the player on the go.

Another interesting approach is using odor as a biometric to identify individuals [21]. In this approach, the tiny quantities of molecules that constantly evaporate and produce the smell, known as odorants, are detected by a special sensor called *e-nose*. *e-nose* is a chemical sensor that can be used to collect unique data about each individual participant. The data can be used to train classification models and to authenticate users [22]. *e-nose* is a rapid, noninvasive, and intelligent online instrument based on the feasibility and effectiveness of odor recognition. Made up of an array of sensors, it is an appropriate pattern recognition system, which is capable of identifying particular smells.

Facial recognition and emotion detection have been used in many applications to classify users. Gabor wavelets is a method to extract features from an image for recognition. For example, analyzing facial images for face recognition by preprocessing or normalizing the face image [23]. As a common rule, the eyes and the

Behavioral Biometrics	Purpose
Keystroke Dynamics	To recognize a person using keystroke dynamics [11].
Keystroke and Mouse Dynamics	Identity theft issues by verifying users based on their keystroke dynamics and mouse activities [26]
Touch and hold a device	Emotion detection from touch interactions during text entry on smartphones [12]
Touch Patterns	continuous authentication on mobile banking applications [19]
Mouse Dynamics	Computer user recognition based on the way a user uses his/her mouse [15]
GAIT	Authorization process based on style or manner of walking [17, 18]
Strategy	Player profile is used to authorize the player on the go [20]
Odor	Human recognition through the odor authentication [21]
Gabor wavelets	To extract features from an image for recognition [23]
Handwriting Biometric	A process of transforming a language represented in its spatial form of graphical marks into its symbolic representation [24]
Speech	Useful for biometric authentication, forensics, security, speech recognition, and speaker diarization [25]

Table 1.
Behavioral biometric research work.

mouth will always be aligned roughly at the same position in same-sized images for face processing. Gabor filters for different scales at different orientations are applied to each facial image for the purpose of creating feature vectors to train machine learning models.

Several researchers considered handwriting biometrics as behavioral biometrics as they are based on actions performed by a specific subject. Handwriting recognition is the task of transforming a language represented in its spatial form of graphical marks into its symbolic representation [24].

Voice recognition is one of the behavioral biometrics that can be used to identify a vocal pattern based on sound variations that are most common in a person's speech. Both speaker identification and speaker verification can be done by capturing important narrow-band speaker characteristics such as pitch and formats [25]. This technique is used for biometric authentication, forensics, security, speech recognition, and speaker diarization.

A brief list of previous studies is given in **Table 1**.

4. Behavioral biometrics solutions in the industry

Not only researchers, but many industry experts are working diligently to improve the applications and performance of behavioral biometric solutions.

4.1 BioCatch

Founded in 2011, BioCatch is working diligently to address next-generation digital identity challenges by focusing on online user behavior. BioCatch has developed several solutions that could improve security in the following use cases: 1) Account opening protection, 2) Account takeover protection, 3) Social engineering scam detection, 4) PSD2 strong customer authentication, etc. [27]. As per BioCatch, "In our digital world, behavior tells all" [27]. Regardless of an attacker's chosen mode of operation, user behavior can never be stolen, spoofed, or replicated. BioCatch has developed solutions that can continuously monitor a user's physical and cognitive digital behaviors. These solutions can be used to analyze thousands of interactions per session and build models to distinguish between genuine and non-genuine users. The solutions are used for several surveillance systems like account opening protection, account takeover protection, advance social engineering, payment scams, proactive mule detection etc.

BioCatch is providing its software products to many leading banks and helping them to prevent identity thefts and other frauds detection and protection. Some major clients for BioCatch are HSBC, American Express, etc. [27].

4.2 Simprints

Simprints works on the motto of "Transforms the way the world fights with poverty". They are working on building technologies that can be used to identify the person with fingerprints to generate biometric ID for data analysis. The plan is to build a technology that can radically increase transparency and effectiveness in global development, making sure that every vaccine, every dollar, every public good reaches the people who need them the most [28].

4.3 PluriLock

Founded in 2016, Plurilock is working to provide an advanced authentication system using behavioral biometrics [29]. They use the concept of device-based gestures to authenticate users using keystroke dynamics and mouse movements in their two products namely, PLURILOCK AWARE and PLURILOCK DEFEND [29].

- Plurilock Aware—deals with the problem of login credentials, and ends up the frustration of typing passwords and OTP. It provides identity verification by recognizing the typing patterns of the users. It is invisible to the users, not-stealable, and takes care of privacy.

- Plurilock Defend—detects the legit person, while the session is on, using continuous authentication. It also monitors the session activity. Using continuous keystroke and mouse monitoring, the risk is reflected and the system is alarmed.

The AWARE and DEFEND products use patented algorithms to bring continuous authentication to highly-regulated environments like government, critical infrastructure, financial services, and healthcare.

4.4 TypingDNA

Using keystroke dynamics, TypingDNA provides continuous authentication. Founded in 2016, TypingDNA works on recognizing a person's typing behavior for authorization. The company had launched four products for verification and authentication purposes:

- VERIFY 2FA—a 2-factor authentication product, which has an AI agent, which examines and saves the typing pattern of a user for future verification [30]. The second product is authentication API. It uses four different ways to authenticate the user.

- Login authentication—when the user logs in for the first time, it will register that typing behavior and will use the created profile to verify the user later. When the user types his login credentials next time, the AI will match it with the first enrollment. If more than 90% of the features match, then the user will be authenticated [31].

- ActiveLock—This product is used to restrict the unauthorized access to the company computers using continuous authentication. If any bizarre typing pattern is recognized by the system, it will automatically lock the computer system. Also, if an authorized person forgets to log out of his computer and any unauthorized person tries to access the data, continuous authentication will catch the unusual behavior and will lock the system [32].

- Focus—Based on the typing patterns, this application helps users to recognize what mood they are in and what time of the day they are more productive. This application works as a mood tracker. When the user types anything, it examines the typing behavior and analyzes several features. This includes: when the user is actively engaged in typing, for how long he was typing, the typing speed and the typing volume. The tool uses AI to predict the mood of the user [33].

4.5 ThreatMark

The company provides a complete package to prevent current and future digital fraud since 2015 [34]. ThreatMark is working to prepare solutions for banks to fight fraud, from early threat detection, over behavioral biometrics to transaction risk analysis.

- Anti Fraud Suite (AFS)—Innovative, feature-rich and modular Fraud Detection Solution for Digital Banking and Payments featuring behavioral profiling, including behavioral biometrics, transaction risk analysis and threat detection in one machine learning-based analytics engine.

- Clair—Unique Solution for Online lending, Gaming and other businesses looking to minimize fraud risk and/or credit risk. Clair is using behavioral profiling and biometrics to identify users, predict future business outcomes, fraud and more.

4.6 3Divi

Founded in 2011, 3DiVi Inc. is an AI technology company focused on the application of deep learning to computer vision [35]. The company is working on developing state-of-the-art API/SDKs that enable smart devices to recognize humans. Their solutions are used by several big companies like Intel, Adidas, LG, Orbbec etc. The company is working hard to enable human-machine interface (HMI) in IoT, smart home, smart retail, smart car, robotics, and digital identity verticals. The product line has several specialized SDKs.

- NUITRACK SDK—a 3D tracking middleware developed by 3DiVi Inc. This is a solution for skeleton tracking and gesture recognition that enables the capabilities of Natural User Interface (NUI) on Android, Windows, and Linux.

- Interactive Android™ Box—Game with gesture recognition—Ultimate platform to build and sell applications with full body and face interactivity.

- Face SDK—face recognition with a suite of solutions designed to enhance business capabilities, automate tasks, and increase overall community safety.

- SEEMETRIX—Anonymous Face Analytics. This solution can be used to detect gender, age, emotions in a fraction of second

4.7 Zighra

Zighra makes authentication more secure than static MFA and enables passwordless experiences [36]. Their platforms, combine insights from generative behavioral models and biological systems to train faster, dynamically adapt, and accelerate execution compared to AI approaches commonly used today.

The software provides task-based authentication where users are asked to perform a specific action as an authenticator to determine whether the user or a bot is trying to use the device, such as holding the phone and swiping across the screen. It also provides security intelligence, using the unique ways a user types, swipes, and taps.

Transaction risk assessment is done using machine learning and behavioral biometrics to ensure the identity of the user on the device and also provides proof of presence using AI, behavioral biometrics, sensor analytics, and network intelligence together to actively authenticate the identity of the on-device user [36].

They have been awarded an innovation contract to pilot continuous authentication for remote access using patented next generation AI technology by the government of Canada [37].

4.8 VoiSentry

A speaker identification and verification (ID&V) system developed by Aculab, that captures tens of thousands of unique voices and speech characteristics to authorize the user on the go [38]. This solution is an ideal system for voice biometric authentication system in terms of performance and accuracy.

4.9 Cynet

Cynet's user behavior analytics system continuously monitors and profiles the user activity [39]. This profile is later used to define a legitimate behavioral baseline and identify anomalous activity to indicate any compromise in the user accounts. It provides real-time monitoring of all the interactions from the time users initiate by logging in.

4.10 BehavioSec Inc.

The BehavioSec solution provides a continuously learning AI subsystem with preweighted machine learning models based on prior analysis, using a hybrid of offline and online calculations [40]. The company leverages APIs, SDKs, and rich behavioral biometrics insights, that can be used to embed seamless security into the existing systems.

4.11 SecureAuth Inc.

Working toward deploying MFA in a digital world [41]. The initiatives are password authentication, portal and web apps security, RSA migration etc. The products are deployed in several industries like healthcare, retail, energy, financial, and public sectors.

4.12 UnifyId

They are the developers of a passive behavioral authentication platform designed to identify users without any conscious user action [42]. The platform developed

Company Name	Year	Types	Used by
BioCatch [27]	2011	Typing speed, Swipe pattern, mouse clicks	HSBC, Itau, BARCLAYS, nab, American Express, citi VENTURES, 86400 banks, NatWest
Simprints [28]	2012	Wireless Fingerprint scanners	BRAC, Cohesu
Plurilock [29]	2016	Keystroke dynamics, Pointer dynamics	US federal agencies
TypingDNA [31]	2016	Keystroke dynamics	Microsoft Azure, ForgeRock, Optimal IdM, BBVA, Proctoru, Capgemini
ThreatMark [34]	2015	Mouse events, keystroke dynamics, site navigation patterns, interaction with website elements	SLOVENSKÁ SPORITEĽ ŇA(Bank), SBERBANK
DiVi [35]	2011	Facial Recognition, Skeleton tracking	Intel, Adidas, LG, Orbbec
Zighra [36]	2010	Task-based authentication using behaviors such as holding the phone and swiping across the screen	Government of Canada innovation Fund
VoiSentry [38]	2018	Speaker identification and verification system	ForgeRock, University of York, MyForce
Cynet [39]	2018	Behavior analytic System to continuous monitoring	Darktrace, Microsoft Azure, Vectra Networks
BehaioSec Inc. [40]	2010	The API can turn behavior into actionable intelligence with just a few lines of code	IDG, Gartner, Goode Intelligence
SecureAuth Inc. [41]	2015	Identity Security Without Compromise	Xerox, Michaels, Unisys
Unify Id	2015	Passive behavioral authentication platform designed to identify users without any conscious user action	US banks
SecureTouch Inc.	2014	Deliver continuous authentication technologies to strengthen security and reduce fraud	Zaraz, Neon Media, TimeRack

Table 2.
Behavioral biometric commercial organizations.

utilizes sensor fusion with machine learning to provide enhanced accuracy while improving the user experience. This helps in authentication both in application and in the physical world.

4.13 SecureTouch Inc.

A pioneer in the field of behavioral biometrics for mobile. They work to deliver continuous authentication technologies to strengthen security and reduce fraud while improving customers' digital experience [43]. Their systems seamlessly collect and analyze a dynamic set of over 100 different behavioral parameters like keyboard-typing, scroll-velocity, touchpressure, and finger size to automatically create a unique user behavioral profile, which can be used for authorization later.

Table 2 provides a summary of the companies working on behavioral biometrics technology.

5. Continuous authentication use cases using behavioral biometrics

With the estimation of the growing behavioral biometrics market which is expected to reach $4.62 USD billion by 2027, it has almost grown in every area of usage [44]. In this section, we present some common applications where behavioral biometrics is used very extensively.

- ***Student authentication using typing biometrics***—the need is to have continuous identity and authorship assurance throughout the learning activities within the existing learning space for learning and assessment [45]. Behavioral biometrics is used to make a model that can be applied to measure the degree of learner collaboration with peers and also define and verify the interaction with the course content. Also helpful in validating authorship of the academic artifacts.

- ***Proliferation in desktop and workplace computing***—Same keyboard, mouse, and touch patterns together can help in authorization and controls on the desktops and workplace computer systems.

- ***Customer Authentication with 2FA, without sacrificing UX***—Behavioral biometrics are the innovative and reliable way to secure customer accounts. Many national and international banks have started considering keystroke dynamics or touch patterns as a person's unique characteristics for their authorizations.

- ***Criminal profiling***—Behavioral biometric is used by police and FBI investigators to determine the personality and identity of the individuals who may have committed a crime based on their behavior exhibited during the criminal act and matching it with the stored profiles of the criminals.

- ***Jury profiling***—A BB technique used by lawyers and prosecutors, which can predict the action of the particular potential juror based on their current behavior, overall physical and psychological appearance.

- ***Plan recognition***—To understand the goals of an intelligent agent by analyzing their observable actions by creating a map of their temporal sequencing.

- ***eHealth and Well-being***—In the health care system, it is possible to make the diagnosis based on how a person behaves. For example, monitoring the way the patient is speaking, typing, talking, or engaging in other daily activities. By comparing it with previous data, it is possible to draw appropriate conclusions about the state of the health of a patient.

- ***Healthcare services***—For patient management and electronic health records, voice biometrics is used to provide an additional layer of security that prevents unauthorized access to patient records.

- ***Avoiding User Carelessness***—It is not very uncommon for a human to make mistakes. Sometimes, mistakes may open the door for malicious intrusions. If this happens, behavioral biometrics will help in quick detection and flagging the intrusion.

- ***License Mismanagement***—Although licenses are personal and individual, still users may use them illegally by sharing or stealing. Behavioral biometrics can be used to eliminate the associated risks by ensuring and verifying that only the named persons are using licensed services or products.

- ***Contact Center authorization***—The most common use of voice biometrics is in the contact center space where it is useful for verifying and authenticating the callers, which in turn saves time and effort for both the customer and the agent.

- ***Preventing account sharing practices***—Many companies are using BB for authentication and fraud prevention purposes. Banking is one of the sectors where behavioral biometrics is now commonly used. It is used to authenticate whether the person using the service is genuine or not.

- ***Workforce authentication***—Possibility is to even identify the workers based on the unique behaviors they have while interacting with the devices. This is possible through a true friction-less and less invasive system that can be built using existing hardware capabilities.

- ***Customer Onboarding***—The behavioral biometrics provide insights that provide the global organizations an actionable intelligence, that can be used to create a secure and frictionless digital customer authorizations.

- ***Online Lending***—Behavioral biometrics combined with machine learning and risk assessment techniques provide a much more innovative approach to online user authentication, which helps the lending organizations to take quick actions and early approvals.

- ***Preventing Online and mobile banking Frauds***—Most of the banks and retailers ate tracking their users' way of typing, swiping, and tapping on the devices to make behavioral biometric profiles for authorizations. This helps in reducing fraud. For example: The Royal Bank of Scotland has done a collection of biometric behavioral data, 2 years ago on private banking accounts for wealthy customers. They are now expanding the system to all of its 18.7 million business and retail accounts, to enhance security and stop all the online frauds.

- ***Cyber Threat Detection***—Monitoring user behavior is one of the best ways to detect cyber attacks and fraud in real-time. It focuses on detecting anomalous user behavior by continuously monitoring and matching it with the profiles recorded in the system.

- ***Access Control Systems***—Behavioral biometric's GAIT analysis can be used for access control systems very effectively. This monitors the walking patterns of a human and access can only be granted to the building quickly on approved authorizations, especially in congested areas.

- ***Endpoint protection***—Behavioral biometrics provides endpoint protection ensuring the whole enterprise to be protected. It enables safe, remote access to the servers, from any end device, used by the workers. Protects both the devices (nodes) as well as the servers.

- ***A critical security component for the IoT.***—Passive continuous re-authentication of the users without notifying them is required in IoT for enhanced security. It may even lock the system automatically in case the user is inactive or irregular or anomalous behaviors are observed by the system.

5.1 Behavioral biometric usage timeline

Behavioral biometrics is totally based on artificial intelligence and machine learning. In the 1960s, Dr. Gunnar Fant and Kenneth Stevens created the first model of speech production using X-rays and then in 1970, Dr. Joseph Perkell used those findings to create a speech recognition biometric model.

A timeline of behavioral biometric solutions is given in **Table 3**.

Year	Biometric Used	Used for
BC–220 AD	Use of handprints as evidence in Qin Dynasty	Crime Investigations
The century	Chinese practice of using fingerprints	Personal Identification
1641–1712	Friction ridge skin observations	Plant Anatomy
1856	Observations on permanence	Identification
1886	Observation on fingerprints for Crime scene investigations and criminal identification	Authorization
40's	Morse code authentication in WWII	Authentication
1942	Telegraph operators unique tapping rhythm	Identification
1949	Iris Patterns	Identification
1959	Computer's ability to learn on its own, without human intervention	Learning

Year	Biometric Used	Used for
1960	First model speech production using X-rays of speaking subjects	Authentication
1960	Facial recognition	Identification
1965	Signature recognition system	Identification
1970	Dynamic signature and fingerprints recognition	Identification
1970	An early form of biometric modeling using full-motion x-rays and the previous work of Drs. Fant and Stevens, even used today	Authentication
1980	Speech Group to promote voice recognition tech	Recognition
1991	Real time face recognition	Recognition
1996	Hand geometry recognition gets deployed at Olympics	Identification
1999	ICAO initiates study on biometrics and MRTD	Issuance and acceptance
2001	Face recognition is deployed at the Super Bowl	Recognition
2001	attacks on the World Trade Center draw attention to the need for continuous authentication as a new security measure in global information systems	Security
2002	DARPA launches Total Information Awareness (TIA), the first large-scale use of technologies designed to mine data sets for identifying biometric information	Identification
2004	US-VISIT (United States Visitor and Immigrant Status Indication Technology) becomes operational	Authorization
2006	innovative new algorithms to rapidly and transparently identify computer users as they work	Continuous authorization
2010	Keystroke Dynamics embedded in consumer products	Authorization
2011	Osama bin Laden's body gets identified with biometrics	Identification
2013	Mobile biometrics	Authorization
Mid 2010s	Continuous authentication for mobile application security	Authorization
Mid 2010s	Biometric systems to improve security as well as the system performance	Authorization
Late 2010s	Electric vehicles with face biometrics	Authorization
2018	World's first phone with under-display fingerprint sensor	Identification

Table 3.
Behavioral biometrics timeline.

6. Conclusion

Behavioral biometrics technologies are promising solutions that are designed to complement and improve systems security that is mainly based on physical biometrics. Behavioral biometrics is based on the analysis of unique parameters such as body movements, keystroke dynamics, and device-based gestures. The very common predictions about behavioral biometrics are its increased adoption for authorization, enhance proactive cyber security, and more accurate anomaly detection.

Behavioral biometrics can bring significant benefits to organizations and users. Moreover, it can also be used in emerging fields, such as improving the security of the internet of things, in addition to several traditional environments.

Although there are several challenges in the development and adoption of behavioral biometric systems, they are becoming more popular solutions as they work seamlessly without user intervention and special hardware requirements. Additionally, since the behavioral biometric system cannot be easily fooled using stolen data as the authentication happens dynamically, it provides increased security and convenience. The biggest challenge that faces the adoption of such systems is the need to constantly retrain their classification models to maintain high accuracy rates. Users' behavioral patterns can change based on many parameters, such as emotions, again, illness, etc. Moreover, if the behavioral biometrics depends on sensors or smart devices to collect raw data, such as in keystroke dynamics, then the classification models need to create a new profile every time a user uses a new device. This constant re-training requires the utilization of advanced machine learning techniques, such as re-enforcement learning. Despite these challenges, behavioral biometrics are becoming more popular every day and the market trends show that they are here to stay.

Author details

Mridula Sharma* and Haytham Elmiligi
Computing Sciences, Thompson Rivers University, Kamloops, BC, Canada

*Address all correspondence to: msharma@tru.ca

References

[1] Singh JP, Jain S, Arora S, Singh UP. A survey of behavioral biometric gait recognition: Current success and future perspectives. Archives of Computational Methods in Engineering. 2021;**28**(1): 107-148

[2] Sultana M, Paul PP, Gavrilova M. A concept of social behavioral biometrics: Motivation, current developments, and future trends. In: 2014 International Conference on Cyberworlds. Cantabria, Spain: IEEE; 2014. pp. 271-278

[3] White paper: Behavioral biometrics. Technical Report MSU-CSE-06-2, International Biometrics Identity Association, 1090 Vermont Avenue, NW 6th Floor Washington, DC 20005, January 2006

[4] Habeeb A. Comparison between physiological and behavioral characteristics of biometric system. Journal of Southwest Jiaotong University. 2019;**54**(6):1-9

[5] Yampolskiy RV, Govindaraju V. Taxonomy of behavioural biometrics. In: Behavioral Biometrics for Human Identification: Intelligent Applications. Hershey: IGI Global; 2010. pp. 1-43

[6] Alsaadi IM. Study on most popular behavioral biometrics, advantages, disadvantages and recent applications: A review. International Journal of Scientific & Technology Research. 2021; **10**(01):15-21

[7] Liu S, Silverman M. A practical guide to biometric security technology. IT Professional. 2001;**3**(1):27-32

[8] Choi M, Lee S, Jo M, Shin JS. Keystroke dynamics-based authentication using unique keypad. Sensors. 2021;**21**(6):2242

[9] El Zein D, Kalakech A. Feature selection for android keystroke dynamics. In: 2018 International Arab Conference on Information Technology (ACIT). Werdanye, Lebanon: IEEE; 2018. pp. 1-6

[10] Halakou F. Feature selection in keystroke dynamics authentication systems. In: International Conference on Computer, Information Technology and Digital Media. Tehran, Iran: Research Gate; 2013

[11] Qi Y, Jia W, Gao S. Emotion recognition based on piezoelectric keystroke dynamics and machine learning. In: 2021 IEEE International Conference on Flexible and Printable Sensors and Systems (FLEPS). Manchester, United Kingdom: IEEE; 2021. pp. 1-4

[12] Ghosh S, Hiware K, Ganguly N, Mitra B, De P. Emotion detection from touch interactions during text entry on smartphones. International Journal of Human-Computer Studies. 2019;**130**: 47-57

[13] Ahmed A, Traore I. Mouse dynamics biometric technology. In: Behavioral Biometrics for Human Identification: Intelligent Applications. Hershey: IGI Global; 2009. pp. 207-223

[14] Antal M, Fejer N, Buza K. SapiMouse: Mouse dynamics-based user authentication using deep feature learning. In: 2021 IEEE 15th International Symposium on Applied Computational Intelligence and Informatics (SACI). Timisoara, Romania: IEEE; 2021

[15] Monaro M, Cannonito E, Gamberini L, Sartori G. Spotting faked 5 stars ratings in e-commerce using mouse

dynamics. Computers in Human Behavior. 2020;**109**:106348

[16] Bhatnagar M, Jain RK, Khairnar NS. A survey on behavioral biometric techniques: Mouse vs keyboard dynamics. International Journal of Computer Applications. 2013;**975**:8887

[17] Chellappa R, Veeraraghavan A, Ramanathan N. Gait Biometrics, Overview. US, Boston, MA: Springer; 2009. pp. 628-633

[18] Elgammal A. Gait Recognition, Motion Analysis for. US, Boston, MA: Springer; 2009. pp. 639-646

[19] Estrela PMAB, Albuquerque RO, Amaral DM, Giozza WF, Nze GDA, de Mendonça FLL. Biotouch: A framework based on behavioral biometrics and location for continuous authentication on mobile banking applications. In: 2020 15th Iberian Conference on Information Systems and Technologies (CISTI). Seville, Spain: IEEE; 2020. pp. 1-6

[20] Yampolskiy R, Govindaraju V. Game playing tactic as a behavioral biometric for human identification. In: Behavioral Biometrics for Human Identification: Intelligent Applications. Hershey: IGI Global; 2010. p. 385

[21] Zhanna Korotkaya. Biometric person authentication odor. Semantic Scholar; 2003:1

[22] Borowik P, Adamowicz L, Tarakowski R, Siwek K, Grzywacz T. Odor detection using an e-nose with a reduced sensor array. Sensors. 2020; **20**(12):3542

[23] Amin MA, Yan H. Gabor wavelets in behavioral biometrics. In: Behavioral Biometrics for Human Identification: Intelligent Applications. Hershey: IGI Global; 2010. pp. 121-150

[24] Plamondon R, Srihari SN. Online and off-line handwriting recognition: A comprehensive survey. IEEE Transactions on Pattern Analysis and Machine Intelligence. 2000;**22**(1): 63-84

[25] Ravanelli M, Bengio Y. Speaker recognition from raw waveform with sincnet. In: IEEE Spoken Language Technology Workshop. Ithaca, New York: IEEE; 2019

[26] Moskovitch R, Feher C, Messerman A, Kirschnick N, Mustafic T, Camtepe A, et al. Identity theft, computers and behavioral biometrics. In: 2009 IEEE International Conference on Intelligence and Security Informatics. Richardson, TX, USA: IEEE; 2009. pp. 155-160

[27] BioCatch. 2021. https://www.biocatch.com [Accessed: December 2021]

[28] Simprints [Online]. 2011. Available from: https://www.simprints.com/ [Accessed: November 29, 2021]

[29] Plurilock [Online]. 2016. Available from: https://www.plurilock.com/

[30] VERIFY 2FA [Online]. 2014. Available from: https://www.typingdna.com/verify [Accessed: December 14, 2021]

[31] TypingDNA [Online]. 2016. Available from: https://www.typingdna.com/ [Accessed: November 10, 2021]

[32] ActiveLock [Online]. 2016. Available from: https://www.typingdna.com/activelockcontinuous-authentication [Accessed: November 30, 2021]

[33] Focus [Online]. 2016. Available from: https://www.typingdna.com/focus [Accessed: November 2, 2021]

[34] ThreatMark [Online]. 2015. Available from: https://www.threatmark.com/whythreatmark/ [Accessed: December 2, 2021]

[35] 3Divi [Online]. 2011. Available from: https://www.3divi.com/ [Accessed: November 29, 2021]

[36] Zighra. Zighara Smart Identity Defense [Online]. Ottawa, ON, Canada: Zighra; 2010. Available from: https://zighra.com/ [Accessed: December 2, 2021]

[37] Deepak Dutt. Government of Canada Awards Innovation Contract to Zighra to Pilot Continuous Authentication for Remote Access using Patented Next Generation AI Technology, OTTAWA, ON; 2021. Available from: PRNewswire.com

[38] Aculab. VoiSentry: Easily add Speaker Verification and Authentication to your Applications. UK & USA: Aculab; 2018 Available from: https://www.aculab.com/

[39] Cynet. Monitor User Behavior to Discover Compromised Identities [Online]. 2018. Available from: https://www.cynet.com/platform/threatprotection/uba-user-behavioranalytics/ [Accessed: November 29, 2021]

[40] Burkhard Stiller, Thomas Bocek, Fabio Hecht, Guilherme Machado, Peter Racz, and Martin Waldburger. Protect Users Without Frustrating Them Using AI-Driven Behavioral Biometrics. Technical report, 01 2010

[41] SecureAuth. Identity Security Without Compromise [Online]. 2014. Available from: https://www.secureauth.com/ [Accessed: December 14, 2021]

[42] UnifyID - authtication, reinvented [Online]. 2015. Available from: https://unify.id/index.html [Accessed: December 14, 2021]

[43] Securetouch [Online]. 2016. Available from: https://craft.co/securedtouch [Accessed: November 29, 2021]

[44] Grand View Research. Press Release: Behavioral Biometrics Market Size Worth $4.62 Billion by 2027. San Francisco, United States: Grand View Research; 2020 Available from: https://www.grandviewresearch.com/

[45] Amigud A, Arnedo-Moreno J, Daradoumis T, Guerrero A-E. A behavioral biometrics based and machine learning aided framework for academic integrity in e-assessment. In: 2016 International Conference on Intelligent Networking and Collaborative Systems (INCOS). Ostrava, Czech Republic: IEEE; 2016. pp. 255-262

Chapter 5

Biometrics of Aquatic Animals

Mahmoud M.S. Farrag

Abstract

This chapter is a part of the book "Recent advances in biometrics" introduces the importance of biometrics in the aquatic studies in brief view. Biometric measurements (Morphometric, meristics and description) are widely used in various fields' "taxonomy, species identifications, monitoring of pollution, species abnormalities, comparison, environmental changes, growth variation, feeding behavior, ecological strategies, stock management, and water quality of aquaculture. These data were collected from several articles and books of aquatic animals and presented both applications and required considerations for biometric implementations. It is important also to detect sexual dimorphism, adaptations during evolutionary time and diminishing intraspecific competition by increasing niche portioning. The biometrics could be applied for various aquatic organisms as dolphins, sharks, rays, mollusca, crustaceans, protozoa, ... etc. and for specific organs like teeth, otolith and appendages by different techniques and preservations. Scientists are still applying these measurements even with the presence of advanced techniques like PCR as they are low in cost, faster and more applicable. This chapter also presented some recent trends including animal's biometric recognition systems, followed by challenges and considerations for the biometrics implementations. It is recommended to apply biometrics in wide range together with modern techniques considering the specificity of its quality and preservation status.

Keywords: biometrics, importance, applications, aquatic sciences, considerations

1. Introduction

The biometrics is a Greek word divided in to two parts "bio" means life and "metrics" means measurements. Biometric science is an old science concerning the documentation of the features or bio measurements or identification characteristics of the targets which could be human, animals and even fossils. It has been used to describe and record the measurement and biological data for, both animal and human (tracking of the similarities of life forms). It is based on anatomic uniqueness of an individual and specificity of physiological and behavior characteristics. Biometrics approach based on behavior characteristics is less expensive and less dangerous for the user; while physiological approach offers highly exact of identification. However, both kinds provide high level of identification than others like passwords and cards.

In general, biometrics were applied in different platforms [1] as follows.

- Criminalistics (using of biometric identifiers to recognize victims, unknown body and prevent kidnapping for identified children).

- Marketing (using of biometrics to identify owners of loyal cards)
- Time accounting systems at work, schools, etc.
- Security systems (to control the access to the rooms and the internet resources)
- Voting system (to identify/authenticate a person who takes a part in voting during the functionality of voting system).
- It used as apart in passport informations as an international required by various organizations such as demands ICAO standards which involve biometrics in passport.
- Biometrics identifiers are used also for registration of immigrants and foreign workers among immigration Affairs. It allows identifying people even without documents.

In animal, a biometric identifier or measurable could be found as robust and distinctive physical, anatomical or molecular trait that can be used to uniquely identify or verify the claimed identity of an animal [2]. Among the advantages of biometrics usage, it does not cause pain or change in the appearance of organisms. For this reason and others, this chapter focuses on the biometrics in animals particularly aquatic organisms. Their analysis can be considered as a first step to investigate the stock structure with large population sizes. The morphological differentiation of partially-isolated stocks due to environmental differences in the habitats could be known as phenotypic markers [3]. The interactive effects of environment produce morphometric differences within a species, variability in growth, development, and maturation creating a variety of body shapes within a species [4–6]. Hence, it is necessary to identify specimens correctly and investigate other biological traits as growth, mortality, fecundity, trophic relation, parasite relationship, historical and paleontological events [7]. The biometric measurements could be applied on different aquatic organisms as sharks, Rays, Mollusca, Crustaceans, Protozoa, etc. and even for different organs like teeth, otolith and appendages.

It is well known that morphology is directly related to species life history and habitat use. Thus, fish morphometric analysis represents an important tool to determine their systematic, growth variation, population parameters and environmental relationships [8–10]. It also, covers several fields of research such as: ecomorphology evaluating the role of environmental pressures on shaping species diet, feeding behavior, ecological strategies, niche partitioning, habitat use and trophic structure population ecology and metapopulations studies, investigating differences in body shape among populations spatially isolated [9–12]. In addition to that, males and females of the same species may be identified as different species because the intraspecific characteristics, therefore information about morphological sexual variation is important to avoid species misleading identification [13–16]. Moreover, the sexual dimorphism is an important evolutionary adaptation mechanism, and to diminishing intraspecific competition by increasing niche portioning [16, 17]. It establishes the relationship between morphology and behavior, elucidating possible ontogenetic niche shifts and the evolutionary plasticity of an organism [18].

Many biologists and taxonomists are still studying the external biometrics (morphometric and meristics) of the organisms in various research fields, even with the presence of molecular biology techniques, giving faster, and low-cost results [19–23].

From another view, the species identification and population discrimination are important in the biodiversity conservation, natural resources, and fisheries management. In certain cases, particularly when we lost some biometric characters for species identification due to sampling and handling processes, we need intensive measurements. So, the modern morphometric technique needed to be applied; as truss network technique (**Figure 1**); it is applied to provide supplementary taxonomic information to enhance the species identification. It could be used also in case of unclear diagnostic characters available for the identification of species as in ariids species which have overlapped characteristics among several species. This technique was provided by Turan et al. [24] and Abdurahman et al. [25]. In addition, the implementation of biometrics could be applied on the internal parasite and used as species identification of host and as a sexual dimorphism indicator [26], the later author studied the impact of Sacculina sp. parasites, Rhizocephalans (Sacculinidae) on two host crabs *Leptodius exaratus* and *Actaea hirsutissima* in Egypt. Over few years, animal biometrics has become an emerging area of research in computer vision and animal cognitive science [27]. The progress has pointed recognition and modeling systems for the animal biometrics, they are being demonstrated in the real-time applications and applicability for representing and detecting the phenotypic appearance of species, visual features, individuals, behaviors, and morphological characteristics of species [28]. These methods can provide better efforts for designing of emerging algorithms, frameworks, and systems for identification and representation of appearances of species in the emerging field of animal biometrics [28, 29]. Beside the importance of biometrics that has been presented above, the present chapter introduces some applications of biometrics in aquatic animals with the considerations for applying biometrics.

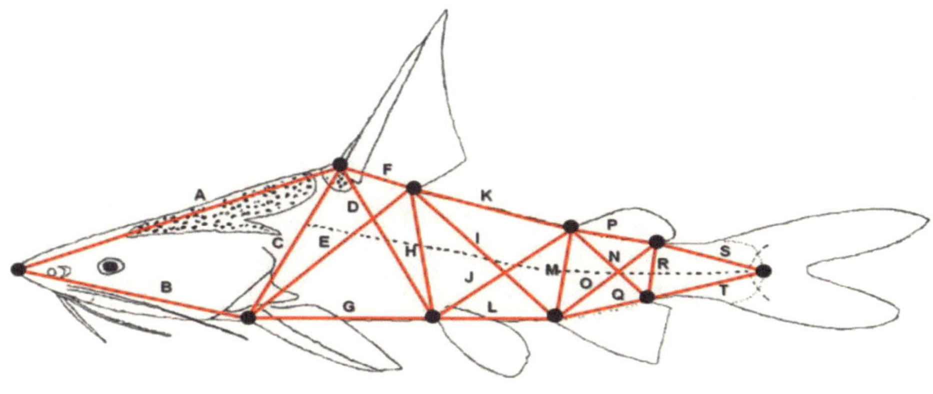

Figure 1.
Truss network distances of ariids family. A: snout to first dorsal fin; B: snout to pectoral fin; C: pectoral fin to F. dorsal fin; D: origin of dorsal fin to pelvic fin; E: pectoral fin to end of dorsal fin; F: origin of dorsal fin to E. dorsal fin; G: pectoral fin to pelvic fin; H: end of dorsal fin to pelvic fin; I: end of dorsal fin to F. anal fin; J: pelvic fin to F. adipose fin; K: end of dorsal fin to F. adipose fin; L: pelvic fin to F. anal fin; M: first of adipose fin to F. anal fin; N: first of adipose fin to E. anal fin; O: anal fin to E. adipose fin; P: length of adipose fin; Q: length of anal fin; R: end of adipose fin to E. anal fin; S: end of adipose fin to caudal fin; and T: end of anal fin to caudal fin.

2. Some of biometric implementations

2.1 Sexual dimorphism

Sexual dimorphism is an important to distinguish males and females. Paiva et al. [30] studied the ontogenetic sexual dimorphism of *Genidens genidens* from the Guanabara Bay, Brazil by applying the morphometric measurements (12 body measurements), for different sexes and different maturity stages. Pearson's linear correlation revealed a significant positive correlation between total length and all other body measures, except for base adipose fin, mouth depth and eye depth for immature females. A significant difference between maturity stages for each sex, indicating a variation in morphometric characteristics driven by sexual dimorphism. Moreover, the differences among all maturity stages indicated an ontogenetic morphological

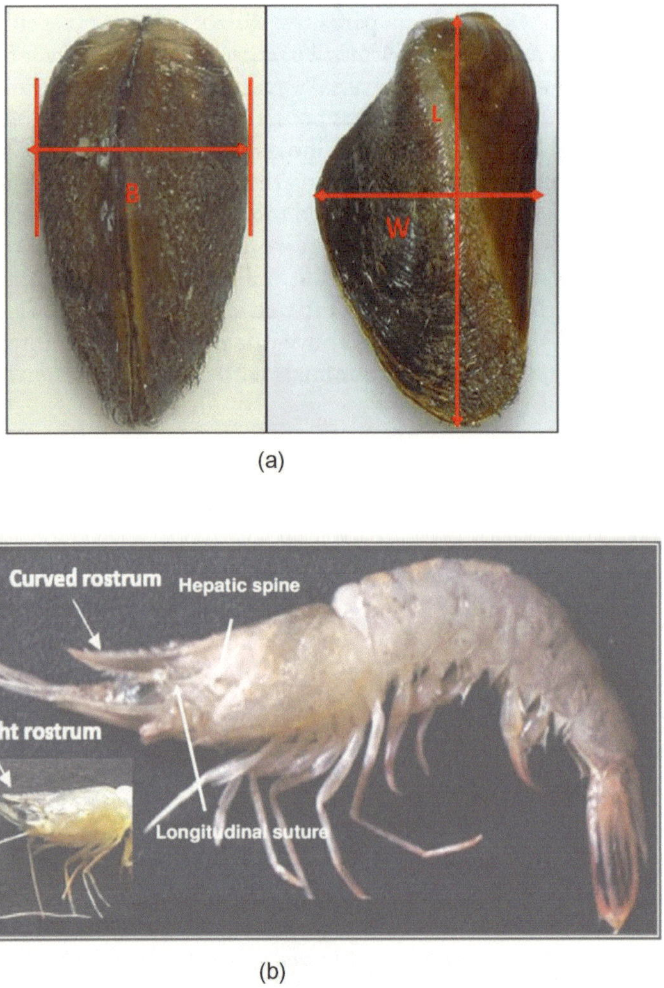

Figure 2.
a. Photographs showing M. auriculatus *morphometeric measurement. L (length), W (width), and B (bottom) [31]. b. Lateral view photograph of fresh adult male and female of southern rough shrimp* T. curvirostris *(Stimpson, 1860) [32].*

difference that started in mature individuals only. The morphometric measurements were applied also on Mollusca (*Modiolus auriculatus*) from the Red Sea, Egypt [31]. The length measurements were applied seasonally and according to sex, to evaluate the changes in sex and growth. (**Figure 2a**). The use of biometrics also was applied by Sharawy et al. [32] on shrimp *T. curvirostris* (Stimpson, 1860) to differentiate males and females (**Figure 2b**).

In this example, it is another application of biometrics on other category of biota "cephalopoda," the morphometric characters of male and female *Sepia pharaonis* from Suez Gulf, Egypt were applied for the whole animal and some internal parts. (**Figures 3a–c**) [33].

The another biometric differentiation was used also for sexual dimorphism of three carangid species (*Carangoides ferdau*, *Carangoides malabaricus*, and *Gnathanodon speciosus*) from the Red Sea, Egypt [16]. The basic statistics of the morphometric indices

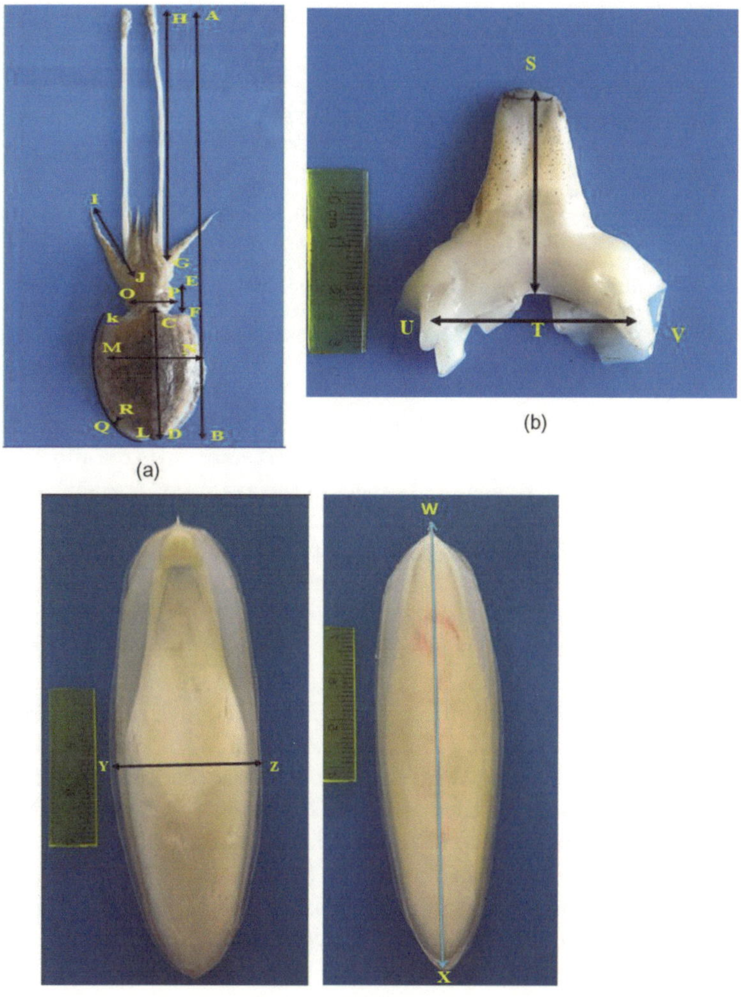

Figure 3.
The different morphometric measurements of S. pharaonis *a; body dorsal view, b: funnel and c: cuttlebone (dorsal view and ventral view).*

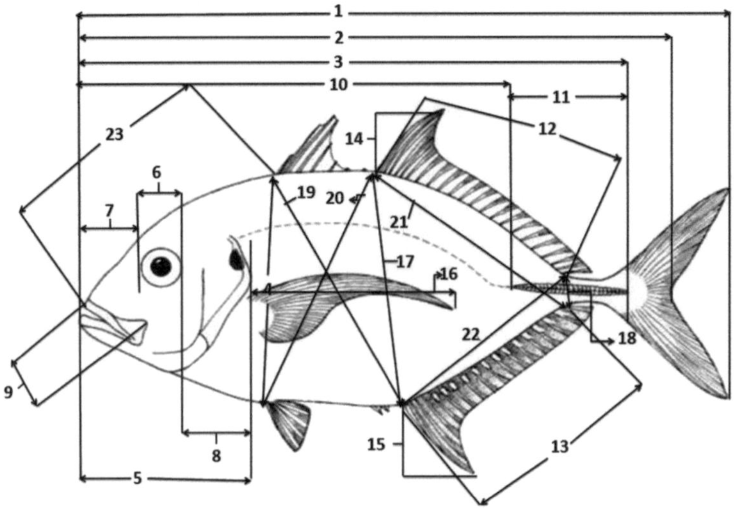

Figure 4.
Schematic illustration of measurements taken on the body of the Three Carangidae Specie considered from the southern Red Sea, Hurghada, Egypt. 1. total length (TL); 2. fork length (FL); 3. standard length (SL); 4. body depth (BD); 5. head length (HL); 6. eye diameter (EyD); 7. snout length (SnL); 8. postorbital length (POL); 9. upper jaw length (UJL); 10. curved lateral line segment length (CLL);11. straight lateral line segment length (SLL); 12. soft dorsal fin base length (SDFL); 13. soft anal fin base length (SAFL); 14. soft dorsal fin height (SDFH); 15. soft anal fin height (SAFH); 16. pectoral fin length (PFL); 17. distance between the first soft dorsal fin ray and the first soft anal fin ray (SDSAFL); 18. distance between anal and dorsal fin insertions (ADFEL); 19. distance between the first spine of the dorsal fin and the first soft anal fin ray (SpDASFL); 20. distance between the first soft dorsal fin ray and ventral fin origin (SDVOFL); 21. distance between the first soft dorsal fin ray and the insertion of anal fin (SDEAFL); 22. distance between the insertion of dorsal fin and the first soft anal fin ray (EDSAFL); 23. predorsal fin length (PRDFL).

(relative to SL or HL) of the three carangid species considered sexual dimorphism (**Figure 4** general diagram) regarding some indices that are size-free and valid as a discriminating tool between males and females of the examined species.

2.2 Light and scanning electron microscopy of internal parasites

The biometric investigations play a role in the field of parasitology and micro examinations. Golemansky and Todorov [34], studied the morphology and biometry of eight marine interstitial testate protozoa, amoebae (*Centropyxiella lucida*, *Cyphoderia littoralis*, *Messemvriella filosa*, *Ogdeniella elegans*, *O. maxima*, *Pomoriella valkanovi*, *Pseudocorythion acutum* and *Rhumbleriella filosa*) by light and scanning electron microscopy. All of them were recognized as a size-monomorphic. By their size frequency distributions, the shell length of *P. acutum* and *O. elegans* were characterized by a not well-expressed main-size class in favor of subsidiary classes, but all species have a shell length ranges in close limits (**Figure 5**).

Another example for using the biometrics, is its application for certain parts like fish scales, since its morphology and ultrastructure characteristics are important for fish identification, taxonomy and phylogeny. The biometrics were applied on the scale morphologically and also on the electron scanning picture of *Acanthopagrus bifasciatus* from the Red Sea, Egypt. A wide spectrum of intraspecific variation between different body regions was recorded in terms of scale morphometric indices and primary and tertiary radii counts. The scale characters including rostral field, outer and inner lateral circuli, grooves, denticles, focus region, granulation in

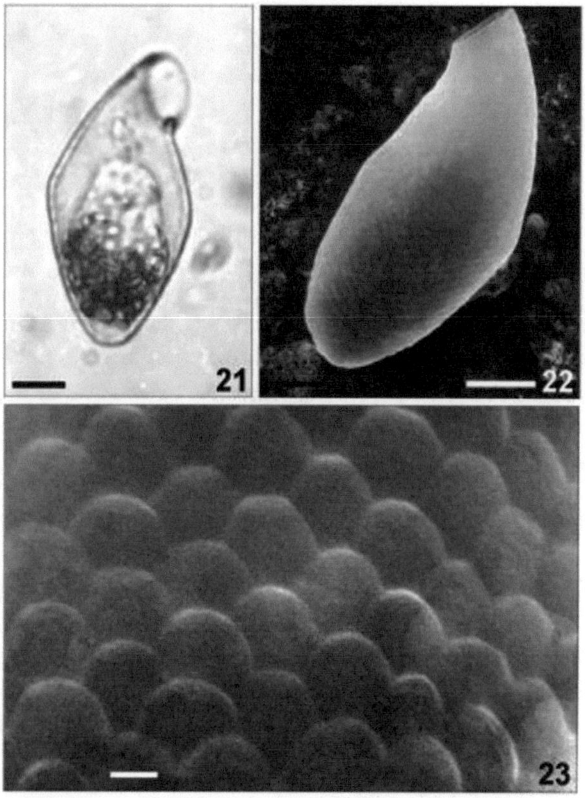

Figure 5.
(21–23). Cyphoderia littoralis. **21**—*lateral view;* **22**—*lateral view;* **23**—*shell structure, showing the imbricated silicious plates (idiosomes) on the shell surface. Scale bars 10 µm (21, 22); 1 µm (23).*

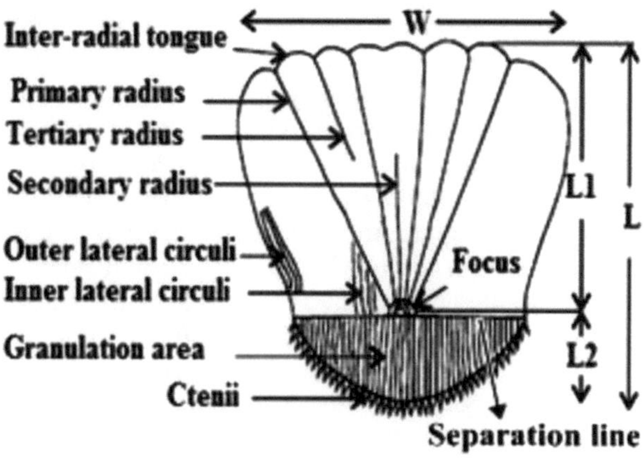

Figure 6.
Schematic drawing scale of A. bifasciatus *showing the different regions, terms and morphometric measurements.*

caudal field and lateral line canal were studied (**Figure 6**) [35]. The ultrastructure by scanning electron microscope (**Figure 7a** and **b**).

Jawad et al. [36] applied the biometric characteristics on another part such as otolith of two species of parrotfish, family Scaridae, from the Red Sea coast of Egypt.

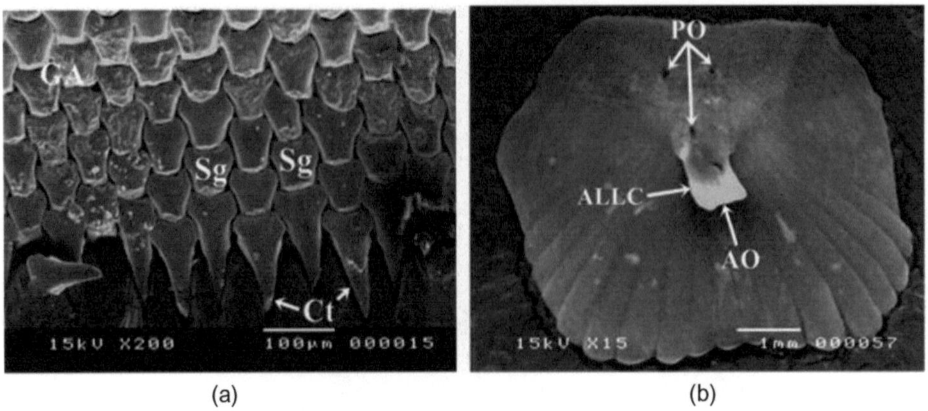

Figure 7.
(a and b): Scanning electron micrographs show scales of A. bifasciatus*; (a): the form of ctenii and segments and (b) the lateral line canal with anterior opening and three posterior opening. Ctenii (Ct), segments (Sg), granulation area (GA), anterior lateral line canal (ALLC), anterior opening (AO) and posterior opening (PO).*

It was applied to identify the most appropriate taxonomic characters that compare or separate these species. Ontogenetic changes in the otoliths of the two scarid fishes become evident. In the otoliths of *Chlorurus sordidus*, *Hipposcarus harid* the characteristics like otolith width, otolith depth, mesial surface shape, lateral surface shape, shape of sulcus acusticus, rostrum and size of rostrum were comparable in small-sized adult fishes, while otoliths of young adults (GI) differed from the adult ones in such characteristics. (**Figure 8a** and **b**).

2.3 Abnormalities in the larval morphology in relation to water quality

The understanding of normal morphology of larvae is very important in aquaculture especially in hatcheries, to evaluate culture conditions for the juveniles and adults. The morphology is an indicator of the abnormalities in the larval morphology in relation to water quality, for production the high-quality individuals [37]. The later author described the allometric growth of Sea bream larvae reared under intensive and extensive conditions, and examined the effect of these conditions on their morphometric proportions; they stated that the intensive marine hatcheries may face many rearing conditions that may reduce the quality of the reared fish, compared to that of the wild ones. These may result in the absence of a swim bladder [38]; osteological and morphological malformations [39], and extra……. The abnormalities in aquatic animals can influence the biometric features, from the modern methods is x-ray utilization, it was applied on three fish species collected from Jubail Vicinity, Saudi Arabia, Arabian Gulf [40] and presented in (**Figure 9**).

2.4 As a comparative key in different habitats

The biometrics were used as comparative tools for species from different habitats and evaluate the effect of environmental conditions. Farrag et al. [20] investigated the biometrics and meristics of puffer fish species *Lagocephalus sceleratus*

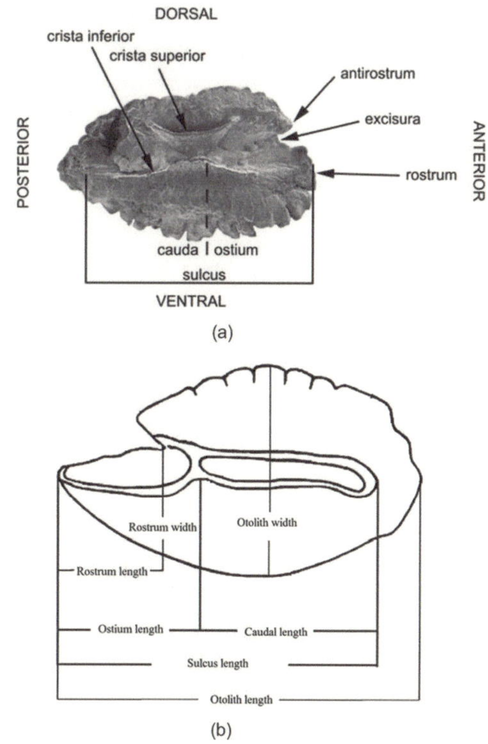

Figure 8.
a. Mesial surface of the left otolith of H. harid *showing its various biometric features. b. Schemeatic diagram of the inner surface of saccular otoliths of a parrotfish showing biometric measurements.*

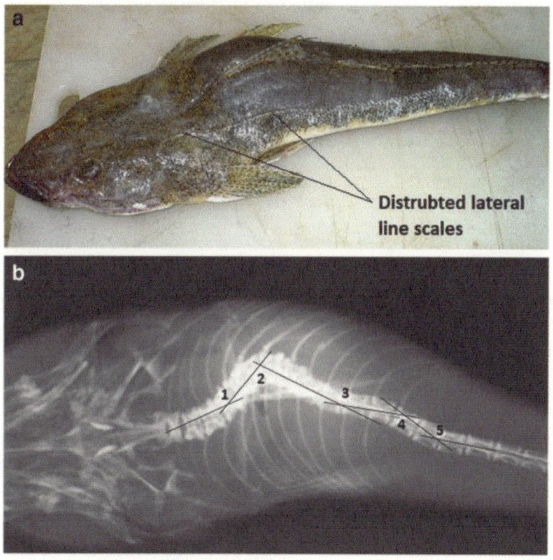

Figure 9.
Platycephalus indicus, 237 mm TL, 231 mm SL, showing scoliosis. a, whole specimen, with deformed lateral line; b, radiograph of Platycephalus indicus, 237 mm TL, 231 mm SL showing skeletal deformity. Numbers 1–5, refers to the angles of curvatures.

from different habitats (Mediterranean Sea and Red Sea, Egypt), Since these characters are sensitive to any environmental changes. The same length range of specimens from both locations was used in morphometric measurements where the resemblance between the sizes of both populations could cause uncertainty with the allometry parameters and it is necessary to avoid size effects. The body width can be strongly influenced by sexual maturation and fullness of stomach as well as the inflating of the body especially for puffer fish that can inflate itself (**Figure 10**).

Using hard parts as spines; the spines are also used in comparison and identification depending on its biometrics and structure. Jawad et al. [13] described structure of the pectoral fins spine of 4 catfish species *Heterobranchus longifilis*, *Clarias gariepinus*, *Chrysichthys auratus*, *Synodontis schall* and *Synodontis serratus* from the River Nile at Asyut City and Lake Nasser, Egypt respectively. The species examined showed variation in the shape of the spines and other biometrics that could be differed among species (**Figures 11** and **12**).

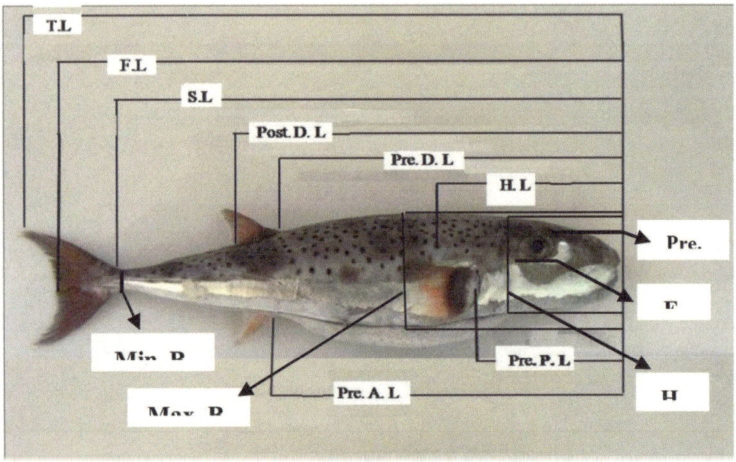

Figure 10.
Lagocephalus sceleratus *showing morphometric measurements.*

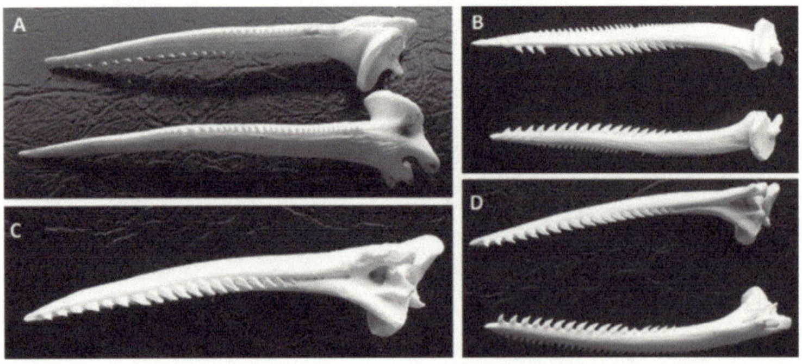

Figure 11.
Left and right pectoral fin spine of Synodontis schall, *400 mm TL (A, C) and* Synodontis serratus, *400 mm TL (B, D) showing dorsal and ventral sides.*

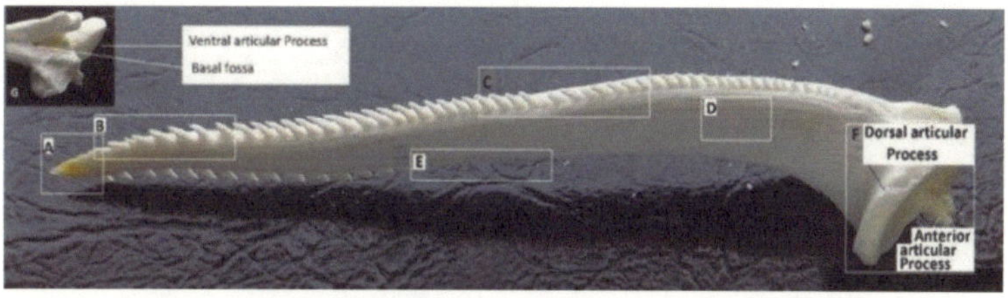

Figure 12.
The enlarged left pectoral-fin spine of S. serratus *(A. anterior distal serrae, B. anterior ridge, C. anterior dentations, D. shaft surface texture of ridges and grooves, E. posterior dentations.*

2.5 As identification and conservatory tools

Biometric methods have therefore been developed to recognize animals based on physical characteristics or behavioral signs. Some of these methods have been used for some time for reliable identification of humans. An animal biometric identifier is any measurable, robust and distinctive physical, anatomical or molecular trait that can be used to uniquely identify or verify the claimed identity of an animal [2].

Sharawy et al. [32] have identified some Penaeid shrimps from Mediterranean, Egypt by different methods. Among them, the authors have applied the biometrics firstly to be correct way to advanced methods or following one. Three penaeid species *Penaeus semisulcatus*, *Metapenaeus monoceros* and *Trachypenaeus curvirostris*. Moreover, they provided the fundamental parameters (**Figures 13a** and **b**) which are important for fisheries management of the currently studied shrimp species. Hence, the conservation resulted after morphological identification has been applied for these species and others others particularly the commercial ones.

2.5.1 Photographing and visual monitoring

The Photographic identification is among biometric methods, it has been used since the 1970s to identify aquatic animals such as dolphins and whales [41]. Individual bottlenose dolphins can be identified by comparing photographs of their fins, which display curves, notches, nicks and tears (**Figure 14**). Whales can be distinguished by the callosity patterns on their heads [42].

The photographing and its treatments using technology used in wide range particularly for wild animals. The most obvious biometric marker is the coat pattern of animals which often appears on major body parts as colourations of either fur, feathers, skin or scales. For example, zebras and tigers can be identified from their stripes; cheetahs and African penguins carry unique spot patterns and snakes have colored rings [28]. From another side, the photographing may face some problem. Problems may occur in the field in different light settings or surroundings, but new techniques including digital photography and video filming have reduced these difficulties. Digital images can also be manipulated to make recognition easier. The method is cheap and at its simplest needs no more than paper and pencil. In addition, observations can be made at a distance, reducing the risk of stress and altered behavior.

2.6 As an indicator of growth

This is another application for morphometric characteristics used to evaluate the growth of species. This was applied on blue swimming crab *Portunus segnis* from the

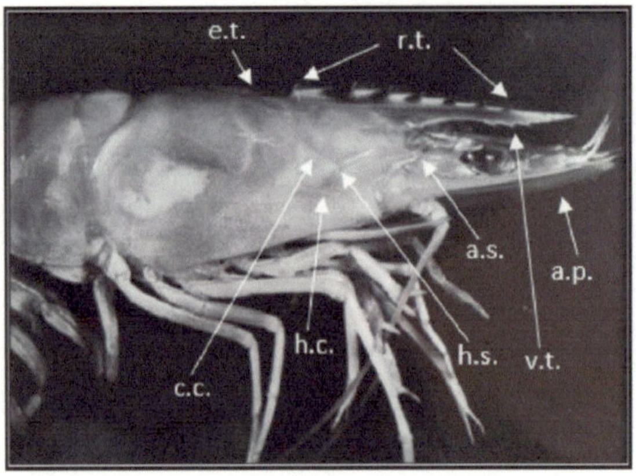

(a)

(b)

Figure 13.
a. Carapace of green tiger prawn P. semisulcatus shows antennal spine (a.s.), antennal peduncle (a.p.), cervical crest (c.c.), epigastric tooth (e.t.), hepatic crest (h.c.), hepatic spine (h.s.), rostal teeth (r.t.) and ventral teeth (v.t.). b. Speckled shrimp M. monoceros shows the antennal peduncle (a.p.), the antennal spine (a.s.), branchiocardiac crest (b.c.), epigastric tooth (e.t.), hepatic spine (h.s.) and rostal teeth (r.t.) together with its dorsal view of telson and tail fan [32].

Figure 14.
Dorsal fins of bottlenose dolphins displaying unique permanent characteristics used for their identification (© 2007 Dolphin Research Center, 58901 Overseas Highway, Grassy Key, FL 33050-6019, USA. http://www.dolphins.org/marineed_photoid.php).

Gulf of Gabes [43]. The carapace width/length- weight relationship was studied in both sexes of crab (**Figure 15**). The exponential values (b) for the carapace width-total weight relationship were distinct between the sexes with a positive growth

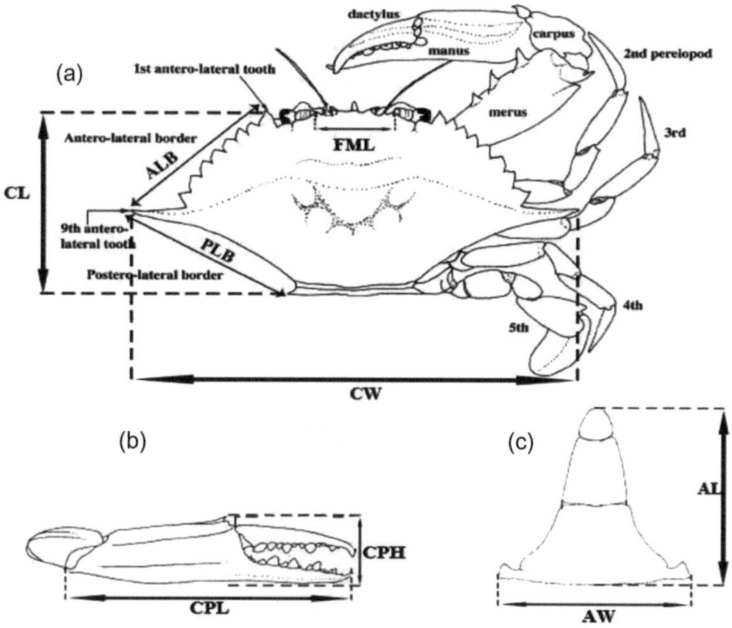

Figure 15.
The morphometric measurements in P. segnis. *a, carapace in dorsal view; b, chela; c, abdomen; CW: Carapace width; CL: Carapace length; ALB: Antero-lateral border; PLB: Postero-lateral border; FML: Frontal margin length; CPL: Chelar propodus length; CPH: Chelar propodus height; AL: Abdomen length; AW: Abdomen width.*

pattern in weight for males, and a negative allometric pattern identified for females. Males were significantly larger and heavier than females, the expected pattern to many crabs.

2.7 Characterization of parasites

The application of the morphology and morphometric was also used to characterize the parasites. It was applied on *Proenenterum* sp. (family: Lepocreadiidae), a new digenetic trematode infecting the pyloric portion of the stomach and the middle part of the intestine of the common sea bream *Pagrus pagrus* fish, they were described by light and scanning electron microscopy for the first time from the coasts of Gulf of Suez and Hurghada city of the Red Sea in Egypt [44]. *Proenenterum* species is characterized by its smaller dimensions and the presence of a large ventral sucker, two lobed testes (**Figure 16**).

2.8 Guidance for computer-cheaper analysis

The biometrics now play an important role in computer analysis of the pictures. The retinal vascular pattern is another biometric trait in animals. The retinal vessels seem to like branching patterns, which are present from birth and do not change during the animal's life. The blood vessels in the eye of each individual can be detected using a retinal scanner. This pattern can be recorded with a hand-held device about the size of a video camera. Some devices can also measure GPS coordinates that used when marking cattle and can be compared to nose-prints. The method is also relatively cheap. Retinal imaging and nose-prints of sheep and cattle were compared by

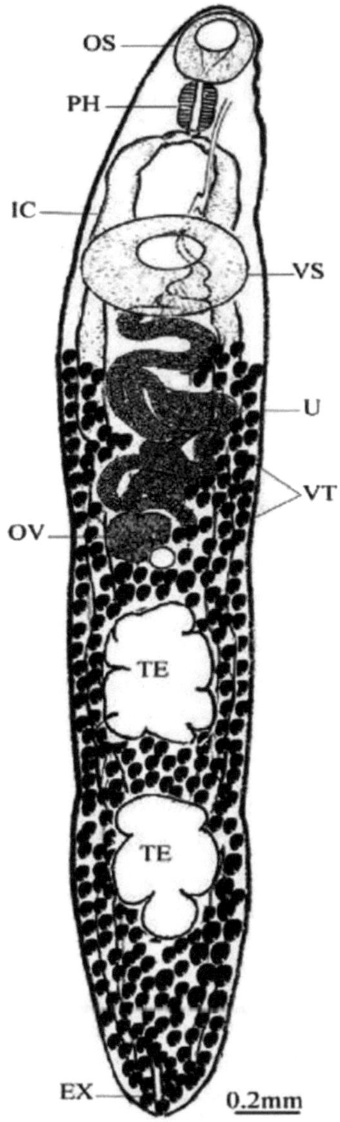

Figure 16.
Line diagram of adult Proenenterum *sp. oral sucker (OS), ventral sucker (VS), pharynx (PH), intestinal caeca (IC), testis (TE), ovary (OV), uterus (U), vitelline follicles (VT), excretory tube (Ex).*

Rusk et al. [45]. However, the nose-prints are a quicker method than retinal scanning, but retinal scans are easy to analyze for inexperienced operators [46]. Computer software for the analysis of digital pictures from both retinal scana and nose-prints makes analysis faster, cheaper and more reliable.

2.9 Movement patterns analysis

The movement pattern is sometimes used as identifier for aquatic animals by analyzing their movement patterns using a tri-axial accelerometry device [47]. By measuring the movements of animals in three dimensions, their movement patterns can be stored and these can be used to diagnose aberrant behavioral patterns, such as those associated with infections. Accelometery may have the potential to be a powerful tool to produce maps for conservation purposes, where animal movements can be plotted.

2.10 Imaging-computerizing treatments

This trend was mentioned by Kumar et al. [48] through recognition systems and this contains different points. For example, the low-Cost Cattle Recognition System Using Multimedia Wireless Network, this system is proposed for verification of individual cattle based on its muzzle point image pattern using wireless multimedia networks. The images are captured using a 20-megapixel camera (system configuration: 14.48 centimeters (5.7-inch) IPS capacitive touchscreen with 1440 × 720 pixels resolution and 283 ppi pixel density, 4GB RAM) and transferred them to the server of cattle recognition using Wi-Fi communication technology. The system performs the image pre-processing on the captured muzzle point image of individual cattle. It mitigates and filter the noise from the captured images and increases the quality [48]. This system could be applied also on the aquatic animals.

The system takes the visual biometric feature characteristics such as coat pattern, body coat pattern, and spot point pattern, and other visual features of species or individual animal. The major issue and challenges of visual animal biometrics-based recognition systems are demonstrated as follows.

- How do species or individual animal gets its body coat pattern? [27].
- What type of suitable algorithms and animal biometrics recognition systems or frameworks is available to compute the visual features from the body coat pattern of species? [27, 28].
- Can detection and representation of visual feature of body pattern of species be possible in their habitats? [28].
- How visual animal biometrics-based recognition and framework can monitor animal population? [28].
- How visual animal biometrics-based recognition system generates unique templates from stored visual biometric feature of species? [27].

3. The considerations that should be taken during the biometric implementation and examples

- Knowing the variations between different organisms and different shapes, therefore should have measurements according to kind of organisms, (Shark, rays, bony fish, crabs, etc).
- It will be better to take the biometric measurements for fresh samples to avoid any error due to preservation or damage in samples. In case of formalin preservation, some changes may happen especially in coloration. So, the more measurements are preferred to be considered.

- In case of comparative study between different habitats, it is preferred to fix the measurements and inputs like length range to avoid bias due to changes in ecological conditions.

- In case of applying biometrics on the internal parts or using scanning techniques, the accuracy, resolution and magnification should be considered.

- In case of using some tools like sensors, it should be easily presented to a sensor and converted into a quantifiable format, should not subjected to changes over time and should differ in the patterns among the general population, the higher the degree of distinctiveness, the more unique is an identifier.

- Biometric methods should not cause pain and do not alter the appearance of the animal, having no effect on the behavior and survivability of the animals, except in some necessary as repeated capture and/or handling.

- In case of visual patterns methods, some species have external characteristics as color, spots, rings, that are easy to recognize and that are specific for each individual. These patterns can be used by photographing using high resolution of digital camera to avoid the problems that may occur in the field in different light settings or surroundings.

- Many common marking procedures also involve tissue damage and therefore cause pain, such as branding (heat, cold or chemicals), tattooing, toe clipping, ear notching and tagging.

- Wearing a mark may alter the animal's appearance, social interaction, other behaviors and ultimately its survival.

- In visual animal biometrics for computer treatment purposes, various issues and challenges lie in coping with unconstrained environment such as variable lighting, partial occlusion of animal body, and extr the captured data sets, images, videos are required to train various computer vision models, framework, and methods.

4. General examples of different aquatic animals and their measurements

The followings are summarized guide for general outer measurements and descriptions that could be taken for various forms and examples of some aquatic organisms including crustaceans, fishes, reptiles and some marine mammals (**Figures 17–24**).

5. Conclusion

In conclusion, the biometrics in organisms (Morphometric, meristics and description) have widely importance used in various fields' "taxonomy, species identifications, monitoring of pollution, species abnormalities, comparison, indicator of environmental changes, growth variation, feeding behavior, ecological strategies, population parameters and water quality of aquaculture operations. The scientists are still applying these measurements even with the presence of advanced techniques

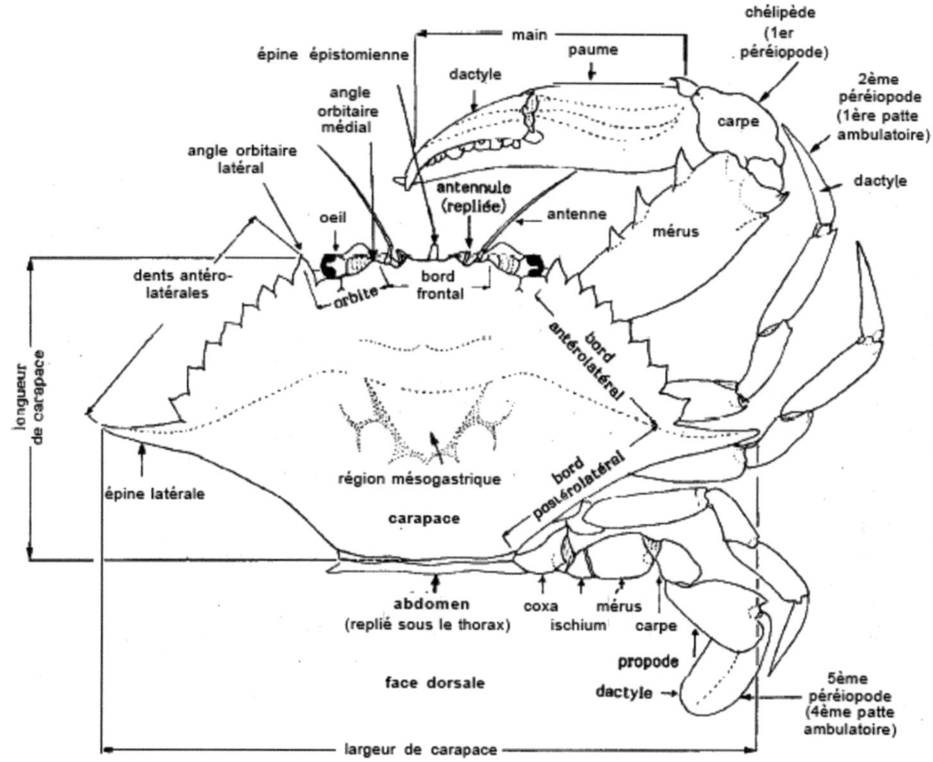

Figure 17.
General morphometric measurement and description of the common form of crabs.

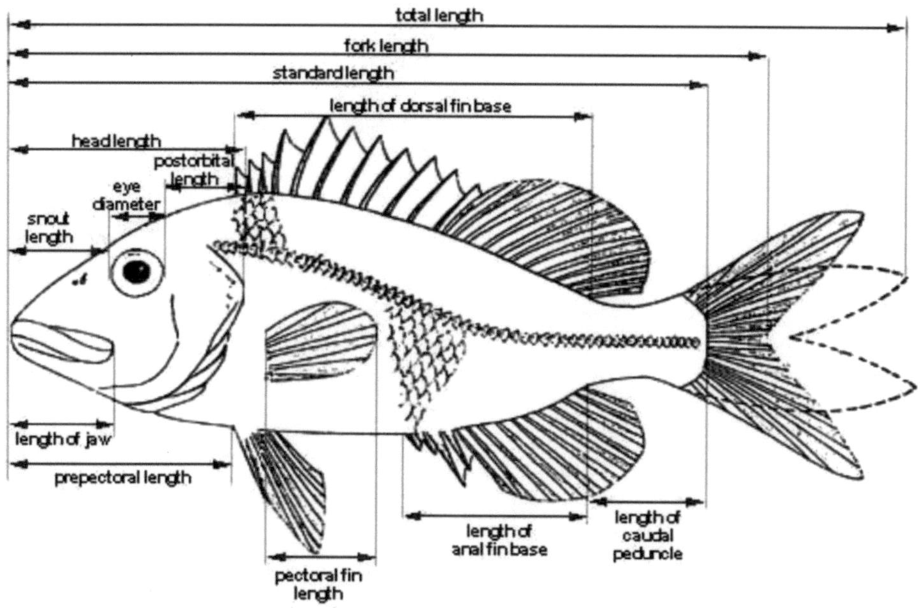

Figure 18.
General morphometric measurement and description of the common form of bony fish.

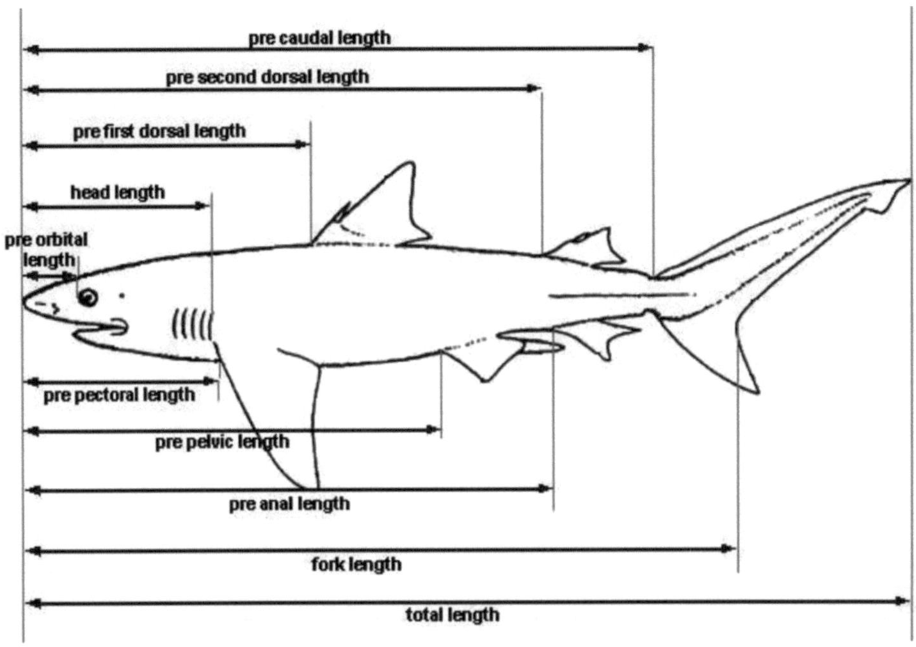

Figure 19.
General morphometric measurement and description of the common form of cartilaginous fish.

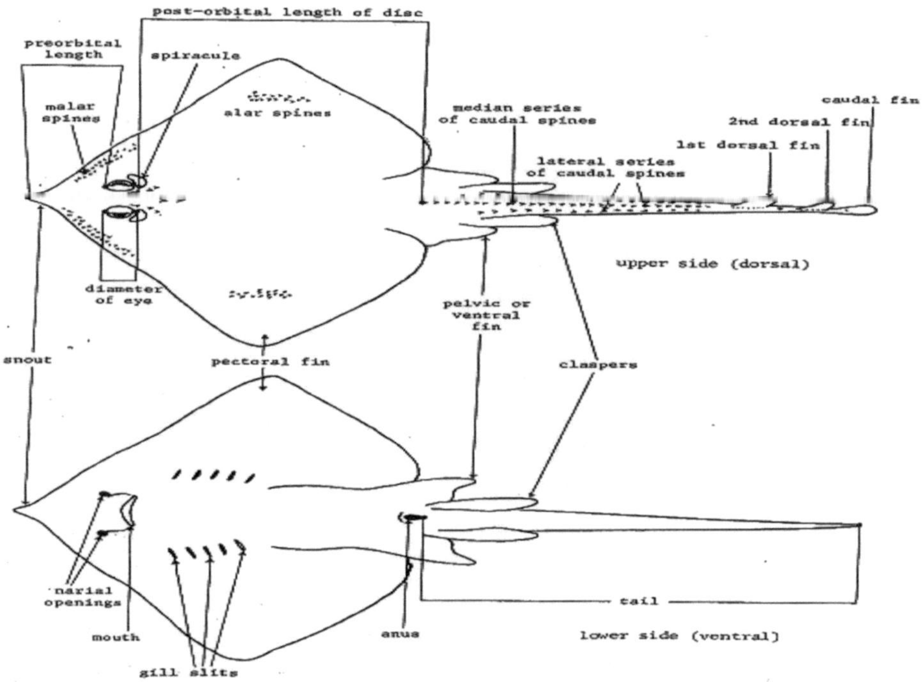

Figure 20.
General morphometric measurement and description of the other form of cartilaginous fish (skates).

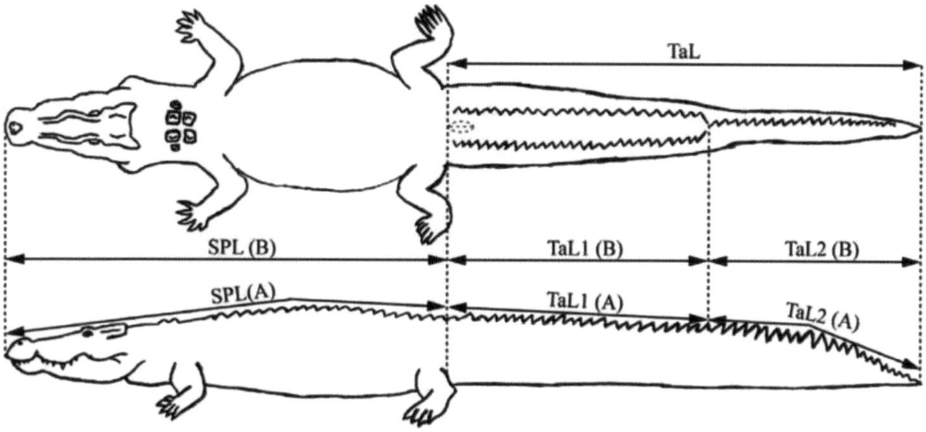

(a)

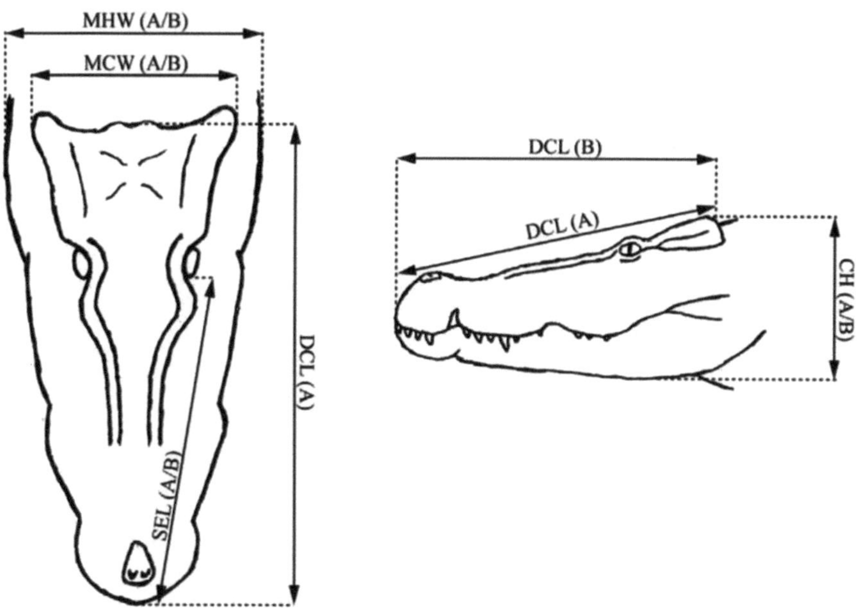

(b)

Figure 21.
Top-down and profile diagrams of entire crocodile (a) and head (b) illustrating measurements taken using Method A (A) and Method B (B). DCL = dorsal cranial length; SEL = snout–eye length; MHW = maximum head width; MCW = maximum cranial width; IOW = inter-orbital width; CH = cranial height; SPL = snout–pelvis length; TaL = tail length; TaL1 = anterior tail length; TaL2 = posterior tail length; SPL + TaL1 = snout–scute junction (SSJ); SPL + TaL1 + TaL2 = total length (TL) [49].

because it is the principal knowledge and first guide, low cost, faster and more available tools used. The considerations for the biometric implementation should be taken during the analysis considering the specificity of the quality, preservation status, kind, form of organism and main target of analysis. Its recommended to give more attention to care the biometrics outer/ inner organisms in scientific studies using the advanced techniques, this will be more beneficially together with other modern techniques which required in certain cases for the same purposes.

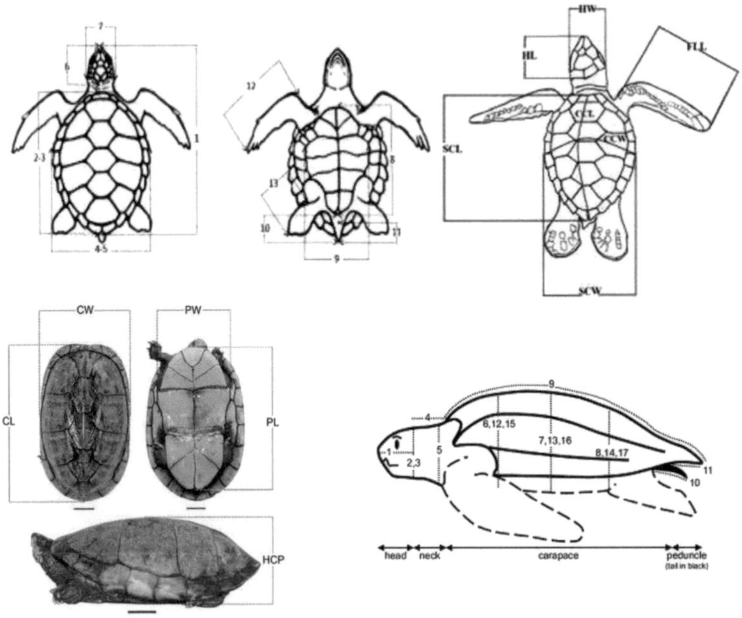

Figure 22.
General morphometric measurement and description of some sea turtles.

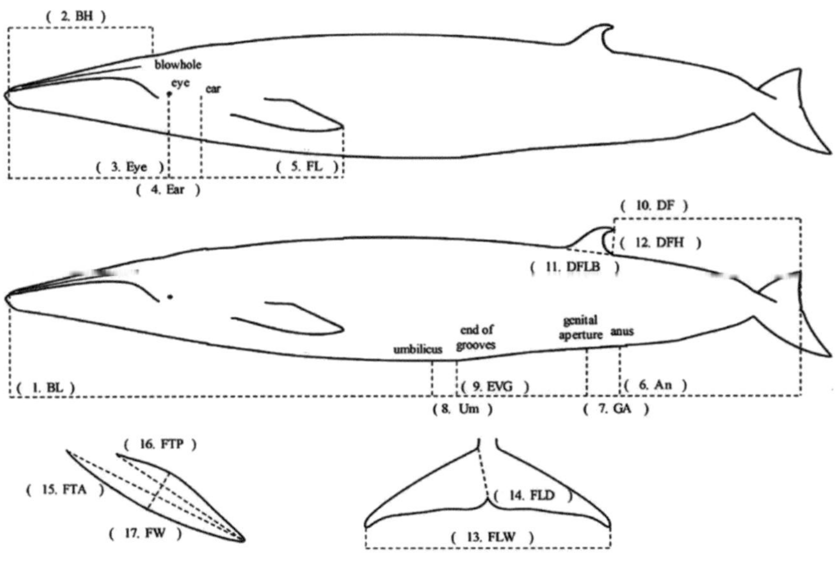

1. BL: Tip of snout to notch of flukes
2. BH: Tip of snout to blowhole
3. Eye: Tip of snout to eye
4. Ear: Tip of snout to ear
5. FL: Tip of snout to tip of flipper
6. An: Notch of flukes to anus
7. GA: Notch of flukes to genital aperture
8. Um: Notch of flukes to umbilicus
9. EVG: Notch of flukes to end of ventral grooves
10. DF: Notch of flukes to tip of dorsal fin
11. DFLB: Dorsal fin, length at base
12. DFH: Dorsal fin, height
13. FLW: Flukes, width tip to tip
14. FLD: Flukes, depth
15. FTA: Flipper, tip to anterior insertion
16. FTP: Flipper, tip to posterior insertion
17. FW: Flipper, maximum width
18. SKL: Tip of premaxilla to occipital condyle
19. SKW: Greatest width of skull

Figure 23.
Measurement points for the body proportions of Bryde's whales. Measurement points were selected based on the study by Mackintosh and Wheeler [50].

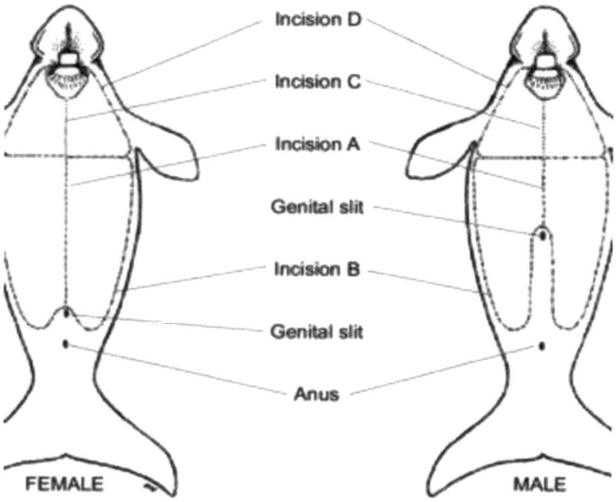

Figure 24.
General morphometric measurement and description of other marine mammals.

Author details

Mahmoud M.S. Farrag
Faculty of Science, Al-Azhar University, Assiut, Egypt

*Address all correspondence to: m_mahrousfarrag@yahoo.com;
mahrousfarrag42@azhar.edu.eg

References

[1] Babich A. Biometric Authentication. Types of biometric identifiers [Bachelor's thesis]. University of Applied Sciences; 2012 53p

[2] Barron UG, Butler F, McDonnell K, Ward S. The end of the identity crisis? Advances in biometric markers for animal identification. Irish Veterinary Journal. 2009;**62**:204-208

[3] Turan C. A note on the examination of morphometric differentiation among fish population: The truss system. Turkey Journal of Zoology. 1999;**23**:259-263

[4] Begg GA, Friedland KD, Pearce JB. Stock identification and its role in stock assessment and fisheries management: An overview. Journal of Fishery Research. 1999;**43**:1-8

[5] ICES (International Council for the Exploration of the Sea). Report of the Study Group on Stock Identification Protocols for Finfish and Shellfish Stocks. ICES C. M. M1; 1996

[6] Pawson MG, Jennings S. A critique of methods for stock identification in marine capture fisheries. Journal of Fishery Research. 1996;**25**:203-217

[7] Ibanez AL, Cowx LG, Higgins PO. Geometric morphometric analysis of fish scales for identifying genera, species, and local population within Mugilidae. Canadian Journal of Fisheries and Aquatic Science. 2007;**64**:1091-1100

[8] Pathak NB, Parikh AN, Mankodi PC. Morphometric analysis of fish population from two different ponds of Vadodara city, Gujarat, India. Journal of Animal Veterinary Fishery Science. 2013;**1**(6):6-9

[9] Sampaio ALA, Pagotto JPA, Goulart E. Relationships between morphology, diet and spatial distribution: Testing the effects of intra and interspecific morphological variations on the patterns of resource use in two Neotropical cichlids. Neotropical Ichthyology. 2013;**11**(2):351-360

[10] Souza MA, Fagundes DC, Leal CG, Pompeu PS. Ecomorphology of *Astyanax* species in estrems with diferente substrates. Zoologia. 2014;**31**(1):42-50

[11] Palmeira LP, Monteiro-Neto C. Ecomor-phology and food habits of teleost fishes *Trachinotus carolinus* (Teleostei: Carangidae) and *Menticirrhus littoralis* (Teleostei: Sciaenidae), inhabiting the surf zone off Niterói, Rio de Janeiro, Brazil. Brazilian Journal of Oceanography. 2010;**58**(4):1-9

[12] Santos BS, Quilang JP. Geometric morphometric analysis of *Arius manillensis* and *Arius dispar* (Siluriformes: Ariidae) populations in Laguna de Bay. Philippine Journal of Science. 2012;**141**(1):1-11

[13] Jawad LA, Farrag MMS, Park JM. Interspecific and intraspecific differences in pectoral-fins spine morphology in Nile River and Lake Nasser catfishes, Siluriformes. Proceedings of the Zoological Institute RAS. 2021;**325**(3):308-322

[14] Marceniuk AP, Menezes NA. Systematics of the family Ariidae (Ostariophysi, Siluriformes), with a redefinition of the genera. Zootaxa. 2007;**1416**:3-126

[15] Mohammad AS, Mustafa AA, Farrag MMS, Osman AGM. Ultrastructure and morphometric of the sagittal otolith as confirmatory identification in three Carangid species,

from the northern Red Sea, Egypt. Aquaculture, Aquarium, Conservation & Legislation—International Journal of the Bioflux Society. 2021;**14**(5):3032-3044

[16] Moustafa AA, Mohamed AS, Farrag MMS, Osman AGM. Sexual dimorphism of morphometrics and meristics for three *Carangidae* species from the Hurghada, Red Sea, Egypt. Iranian Journal Ichthyology. 2021;**8**(3):223-235

[17] Herler J, Kerschbaumer M, Mitteroecker P, Postl L, Sturmbauer C. Sexual dimorphism and population divergence in the Lake Tanganyika cichlid fish genus *Tropheus*. Frontiers in Zoology. 2010;7:4

[18] Galis F, Terlouw A, Osse JWM. The relation between morphology and behaviour during ontoge-netic and evolutionary changes. Journal of Fish Biology. 1994;**45**(A):13-26

[19] Abbas EM, Abdelsalam KM, Geba KM, Ahmed HO, Kato M. Genetic and morphological identification of some crabs from the Gulf of Suez, Northern Red Sea, Egypt. Egyptian Journal of Aquatic Research. 2016;**42**:319-329

[20] Farrag MMS, Soliman TBH, Akel EKA, Elhaweet AAK, Moustafa MA. Molecular phylogeny and biometrics of lessepsian puffer fish *Lagocephalus sceleratus* (Gmelin, 1789) from Mediterranean and Red Seas, Egypt. Egyptian Journal of Aquatic Research. 2015;**41**:323-335. DOI: 10.1016/j.ejar.2015.08.001

[21] Farrag MMS, El-Haweet AAK, Akel EKA, Moustafa MA. Occurrence of puffer fishes (Tetraodontidae) in the eastern Mediterranean, Egyptian coast - filling in the gap. BioInvasions Records. 2016;**5**(1):47-54. DOI: 10.3391/bir.2016.5.1.09

[22] Mekkawy IAA, Mohammed AS. Morphometrics and Meristics of the Three Epinepheline species: Cephalopholis argus (Bloch and Schneider, 1801), *Cephalopholis miniata* (Forskal, 1775) and Vardiola louti (Forskal, 1775) from the Red Sea, Egypt. Journal of Biological Sciences. 2011;**11**(1):10-21

[23] Osman AGM, Farrag MM, Mehanna S, Osman Y. Use of otolithic morphometrics and ultrastructure for comparing between three goatfish species (Family: Mullidae) from the northern Red Sea, Hurghada, Egypt. Iranian Journal of Fisheries Sciences. 2020;**19**(2):814-832. DOI: 10.22092/ijfs.2018.120044

[24] Turan C, Gürlek M, Ergüden D, Yağlioğlu D, Öztürk B. Systematic status of nine mullet species (Mugilidae) in the Mediterranean Sea. Turkish Journal of Fisheries and Aquatic Sciences. 2011;**11**:315-321

[25] Abdurahman SW, Ambak MA, Sherief S, Giat SY, Mohamed AA, ChowdhuryAJK.Morphologicalvariations of *Plicofollis* species (Siluriformes: Ariidae) in Peninsular Malaysia: An insight into truss network approach. Sains Malaysiana. 2016;**45**(1):1-7

[26] Alazaly NAYH. Effect of Sacculina infection on some Crustacean Decapods from the Red Sea, Egypt [Ph.D thesis]. Assiut: Al-Azhar University; 2017

[27] Kühl HS, Burghardt T. Animal biometrics: Quantifying and detecting phenotypic appearance. Trends in Ecology and Evolution. 2013;**28**(7):432-441

[28] Burghardt T. Visual animal biometrics: Automatic detection and individual identification by coat pattern [Doctoral dissertation]. University of Bristol; 2008

[29] Crall JD, Gravish N, Mountcastle AM, Combes SA. Beetag: A low-cost, image-based tracking system for the study of animal behavior and locomotion. PLoS One. 2015;**10**(9):e0136487

[30] Paiva LG, Prestrelo L, Sant'Anna KM, Vianna M. Biometric sexual and ontogenetic dimorphism on the marine catfish Genidens genidens (Siluriformes, Ariidae) in a tropical estuary. Latin American Journal of Aquatic Research. 2015;**43**(5):895-903. DOI: 10.3856/vol43-issue5-fulltext-9

[31] Abdel Razek FA, Ismaiel M, Ameran MAA. Occurrence of the blue crab Callinectes sapidus, Rathbun, 1896, and its fisheries biology in Bardawil Lagoon, Sinai Peninsula, Egypt. Egyptian Journal of Aquatic Research. 2016;**42**:223-229. DOI: 10.1016/j.ejar.2016.04.005 1687-4285

[32] Sharawy ZZ, Abbas EM, Khafage AR, Khallaf AG, Ismail RF, Ahmed HO, et al. Descriptive analysis, DNA barcoding and condition index of Penaeids (Crustacea: Decapoda) from the Egyptian Mediterranean coast. Fisheries Research. 2017;**188**:6 16

[33] Riad R, Atta M, Halim Y, Elebiary N. Taxonomical and morphometric studies on *Sepia pharaonis* Ehrenberg, 1831(Cephalopoda: Sepioidea) from the Suez Gulf (Red Sea), Egypt. International Journal of Environmental Science and Engineering (IJESE). 2016;**7**:11-22

[34] Golemansky V, Todorov M. Shell Morphology, Biometry and Distribution of Some Marine Interstitial Testate Amoebae (Sarcodina: Rhizopoda). Acta Protozoology. 2004;**43**:147-162

[35] Mahmoud UM, El-Gammal FI, Mehanna SF, El-Mahdy SM. Scale characteristics of *Acanthopagrus bifasciatus* (Forsskål, 1775) from the Southern Red Sea, Egypt. International Journal of Fisheries and Aquatic Studies. 2017;**5**(1):417-422

[36] Jawad LA, Hoedemakers K, Ibanez AI, Ahmed YA, Abu El-Regal MA, Mehanna SF. Morphology study of the otoliths of the parrotfish, *Chlorurus sordidus* (Forsskål, 1775) and *Hipposcarus harid* (Forsskål, 1775) from the Red Sea coast of Egypt (Family: Scaridae). Journal of the Marine Biological Association of the United Kingdom. 2017;**2017**:1-10. DOI: 10.1017/S0025315416002034

[37] Koumoundouros G, Kiriakosll Z, Divanach P, Kentourily M. Morphometric relationships as criteria for the evaluation of larval quality of gilthead sea bream. Aquaculture International. 1995;**3**:143-149

[38] Paperna I. Swimbladder and skeletal deformations in hatchery bred Sparus QurQtQ. Journal of Fish Biology. 1978;**12**:109-114

[39] Marino G, Boglione C, Bertolini B, Rossi A, Ferreri F, Cataudella S Observations on development and anomalies in the appendicular skeleton of sea bass, Dicentrarchus lQbr Qx L. 1758, larvae and juveniles. Aquaculture and Fisheries Management. 1993;**24**:445-456

[40] Jawad LA, Ibrahim M, Farrag MMS. Severe scoliosis and fin deformities in three fish species collected from Jubail vicinity, Saudi Arabia, Arabian Gulf. Thalassic: An International Journal of Marine Sciences. 2019. DOI: 10.1007/s41208-019-00145-3

[41] Rugh DJ, Zeh JE, Koski WR, Baraff LS, Miller GW, Shelden KEW. An improved system for scoring photo quality and whale identifiability in aerial

photographs of bowhead whales. Report International of Whale Communication. 1998;**48**:501-512

[42] Wells RS. (2002): Identification methods. pp. 601-608. In: Perrin WF, Wursig B, Thewissen Strauss RE, Bookstein, FL. 1982. The truss: Body form reconstruction in morphometrics. *System Zoology* **31**: 113-135

[43] Hajjej G, Sley A, Jarboui O. Morhometrics and length-weight relationship in blue swimming crab, Portunus sgnis (Decapoda, Brachyura) from the gulf of Gabes, Tunisia. International journal of Engineering and Applied Science (IJEAS). 2016;**3**(12):10-16

[44] Morsy K, Abd El-Monem S, Bashtar A. Morphological and morphometric characterization of a new digenetic trematode, Proenenterum sp.n., infecting the common sea bream Pagrus pagrus from the Red Sea in Egypt. Journal of American Science. 2011;**7**(12):262-267

[45] Rusk CP, Blomeke CR, Balschweid MA, Elliott SJ, Baker D. An evaluation of retinal imaging technology for 4-H beef and sheep identification. Journal of Extensions. 2006;**44**(5) Article 5FEA7

[46] Howell BM, Rusk CP, Blomeke CR, McKee RK, Lemenager RP. Perceptions of retinal imaging technology for verifying the identity of 4-H ruminant animals. Journal of Extension. 2008;**46**(5). Available from: http://www.joe.org/joe/2008october/rb9.php http://www.americanscience.org

[47] Shepard ELC, Wilson RP, Quintana F, Laich AG, Liebsch N, Albareda DA, et al. Identification of animal movement patterns using tri-axial accelometry. Endangered Species Research. 2010;**10**:47-60

[48] Kumar S, Singh SK, Dutta T, Gupta HP. Poster: A real-time cattle recognition system using wireless multimedia networks. In: Proceedings of the 14th Annual International Conference on Mobile Systems, Applications, and Services Companion, ACM, New York, NY, USA: MobiSys '16 Companion; 2016. pp. 48. DOI: 10.1145/2938559.2948871

[49] Britton AC, Whitaker R, Whitaker N. Here be a Dragon: Exceptional size in a saltwater Crocodile (*Crocodylus porosus*) from the Philippines. Herpetological Review. 2012;**43**(4):541-546

[50] Bando T, Nakamura G, Fujise Y, Kato H. Developmental changes in the morphology of Western North Pacific Bryde's Whales (*Balaenoptera edeni*). Open Journal of Animal Sciences. 2017;7:344-355

Chapter 6

MedMetrics: Biometrics Passports in Medical and Clinical Healthcare That Enable AI and Blockchain

Huiqi Yvonne Lu

Abstract

The term biometrics was defined to suggest any measurable biological and biomedical metrics that can be used to identify and confirm the uniqueness of individuals. In this chapter, we would like to introduce an emerging area of biometrics, MedMetrics, that combines patient and drug information managed in coded passports to keep medical information accessible, safe and fraud-resistance. Medmetics includes medical and biological biometrics of patients based on their electronic health records, International Classification of Disease codes, Anatomical Therapeutic Chemical codes, Defined Daily Doses, time-series test results, and personalizsed biological data. By combining the blockchain technology, Medimetrics enables sensitive data sharing in between different clinical settings, allowing monitoring patients' health and care, as well as avoiding identification-related medical mistakes or frauds. MedMetrics Blockchain Passport can be used to identify patients and confirm their previous health conditions without the right of modifying or removing previous records. Medmetrics can revolutionary change the user demographic, shift safety restrictions, define new user applications, and encourage ethical AI regulations in medical science and health care. This chapter will discuss these directions and provide insights into the next generation of biometrics in medical science and health care.

Keywords: MedMetrics, biometrics, personal identity, healthcare, medical health record, NLP, BERT, ICD, blockchain, artificial intelligence

1. Introduction

Biometrics was traditionally referred to as a measurement that can help to identify an individual or to rule out intruders. From using fingerprint with ink on paper for a residence card in ancient China to using fingerprint on the digital optic scanner, from face recognition at door entry system to ECG identification for patients, the biometrics technologies have been re-defined in the last decade to fit the ever-increasing needs in advanced biometrics technologies. The latest developments were derived from the enormous boosting of personal mobile devices, the improved affordability of fingerprint sensors and CCTV cameras, the fast 4G and 5G communication network, and the increasingly successful cases of using biometrics in public and private sectors.

Generally speaking, a typical biometrics identification system will firstly record the physical or behavioural characteristics as the ground truth of personal identity, sometimes combine with personal information, e.g., name, gender, date of birth, address, and identification number if applicable. Secondly, a biometrics system will transform single or multiple biometrics and personal information into one encrypted code representing an individual. There is an algorithm to decide the similarity matrix among different codes. When a user attempts to get authorised by using a biometrics identification system, a decision threshold will be applied to decide whether to grant or deny the authorisation.

There are multiple agents within the organisational network to enable the processes of biometrics authorisation. For example, as shown in **Figure 1**, the Image Acquiring Agent is to collect biometrics information (in this example it is an iris image), the Iris Image Processing Agent is to encoding the biometrics information into a code using an embedded algorithm, and the Iris Image Identification Agent is to compare the attempted code with the code that stored in its database for matching. There are two types of matching in this case. If the biometrics system has confirmed personal information, it is a one-to-one matching. If the biometrics system does not know the personal identification information, it becomes a one-to-many (sometimes one-to-millions of people) identification.

Most mobile and ubiquitous applications, such as mobile face recognition, hospital smartcard-fingerprint system, and hand-writing recognitions, use one-to-one identification. The reason behind it is simple: keep false positives low while having a high specificity rate in the biometrics identification. When biometrics is built on the one-to-many identification, which is very common in public sectors (e.g., when a police officer needs to match identification for suspects), the matching

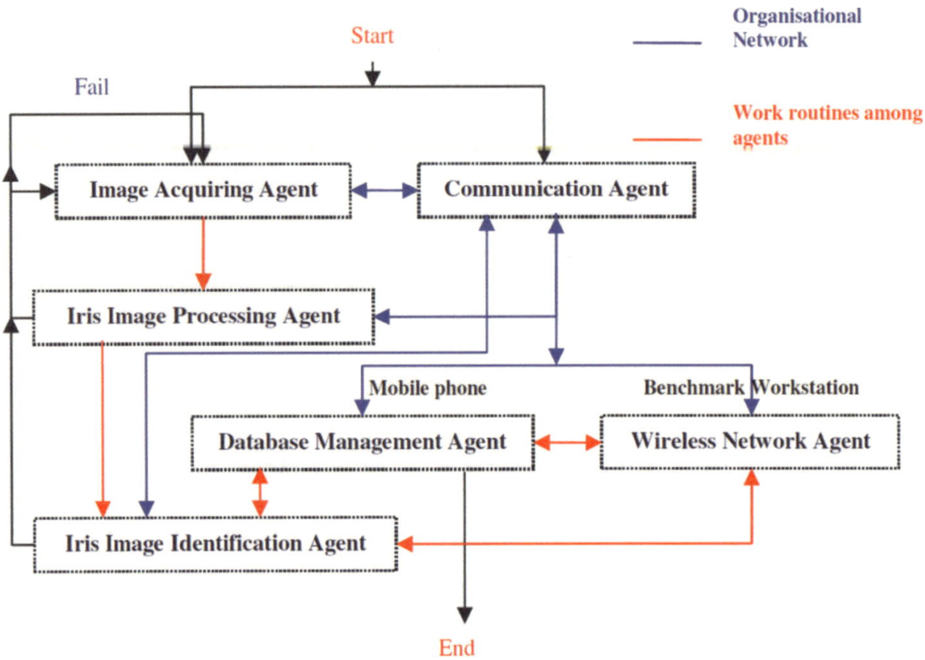

Figure 1.
Distributed agents' organisation network in a typical biometrics system [1].

algorithm and decision threshold become utterly important and need to be customised based on the purpose. A low specificity rate in the biometrics system will lead to a waste of public resources, unfair and biased judgement of individuals.

In this chapter, the author firstly introduces an emerging area of biometrics, namely MedMetrics. The MedMetircs combines medical and biological biometrics with the ultimate aim of identifying patients and their health conditions, drugs, and medical procedures using biometrics codes. To ensure patient data privacy, MedMetrics needs to answer three questions: authentic user, real user, and when to take the action of authorisation. MedMetrics's code is accessible and editable on current events, but all previous information will remain un-editable encryption. It can be used as a fixed baseline blockchain or a time-series blockchain that records temporal codes.

Secondly, this chapter provides insights into MedMetrics in medical science and health care. The author will discuss how traditional biometrics technologies apply to the medical and medical science field, then explain how MedMetrics can revolutionary change the user demographic, shift safety restrictions, define new user applications, and encourage ethical AI regulations in medical science and health care. The novelty of MedMetrics in the areas it applies to, such as the 'MedMetrics Patient Passport' and the 'MedMetrics Drug Passport'. Because the MedMetrics codes are formed based on multiple international standard codes, the introduction of MedMetrics will help ethical and regulatory governance in developing an evaluation of medical biometrics that is trustful and repeatable.

2. MedMetrics in medical science and healthcare

2.1 Recent development in biometrics in healthcare

In the last two decades, biometrics using fingerprints has been accepted and used widely as an identification measure for clinicians in hospital settings. Hospital and healthcare staff can use a smartcard with a user name and password (as a personal signature) and fingerprints for patient data access, test assignment, and drug subscriptions. In the recent 5 years, one encouraging move is that patients started to accept using face recognition on mobile phones to gain healthcare applications. Encrypted face patterns became an acceptable biometric password and inspired several applications. For example, UK residents can now complete the onboarding process for their National Health Service (NHS) login using face recognition via Apple Face ID iProov Genuine Presence Assurance [2]. After the secure facial verification, residents can access essential services online, such as appointments, personal health records, and ordering repeat prescriptions. The Programme Head for NHS login suggested that "More automated tools like this will help us to improve the experience of our users, increase demand capacity and ensure nobody is waiting too long to complete identity verification checks to gain access to their digital healthcare services" [3].

Another application that is emerging is using physiological signs as biometrics. Studies show that electrocardiogram (ECG) signals can be used as a biometrics measure [4–6]. Thanks to the highly individualised nature of the ECG, they are ubiquitous and difficult to counterfeit [7]. However, one of the main challenges in ECG-based biometric development is the lack of large ECG databases.

In this chapter, we propose two novel methods in MedMetrics can play revolution roles in the digital health *Era*. MedMetrics is the medical and healthcare biometrics, which is an ideal tool in health care and clinical settings to assure the correct genuine presence of patients, instruments, drugs, and procedures. The benefits of using MedMetrics identity verification for healthcare include improving digital access to health services, increasing inclusivity and accessibility, enabling contactless engagement, reducing the time and cost of manual administration, protecting privacy, and making security and usability measurable and personalised.

2.2 MedMetrics patient health passport

The first novel method is the '**MedMetrics Patient Passport**'—a unique patient biometrics identity that can be updated with historical information that remains unchangeable. Clinical investigation of medical procedures is highly regulated with national and regional rules and requirements that must be adhered to by investigators, manufacturers, as well as other parties involved in clinical procedures. In the design of 'MedMetrics Patient Passport'. In a digital 'MedMetrics Patient Passport', patients will each have a unique code that includes basic identification, time-series health condition, test results if available, and medical history, including the medication in use.

The basic patient electronic health record (EHR) includes patient name, national identification, ethical group, date of birth, address, etc. This information is personal information biometrics that is often used for patient check-in at hospitals. The time-series health condition will be recorded in the International Classification of Disease (ICD) code and split by hashing space in between events or clinical visits. The medical device will be recorded under the ISO 14155: Clinical investigation of medical devices for human subjects – Good clinical [8].

To evaluate performance in healthcare with MedMetrics, the Healthcare Effectiveness Data and Information Set (HEDIS) can be used. Over 200 million people worldwide enrolled in plans that report HEDIS results. The HEDIS includes more than 90 measures across 6 domains of care, namely: effectiveness of care, access and availability of care, the experience of care, utilisation and risk-adjusted utilisation, health plan descriptive information, and measures reported using electronic clinical data systems [9].

2.3 MedMetrics drug passport

The second method that MedMetrics can help make a revolution in the digital healthcare *Era* is the '**MedMetrics Drug Passport**'. This passport has information about drug's name, place of birth (batch number), place of issue (manufacturer's name), visiting history (known usage in specific health conditions in the format of ICD code), and the record of rejection of entry (known interaction with other drugs, which is recorded in the Anatomical Therapeutic Chemical code). The Anatomical Therapeutic Chemical (ATC) code is a unique code assigned to medicine according to the organ or system it works on. The ATC and the Defined Daily Dose (DDD) as a measuring unit have become the gold standard for international drug utilisation monitoring and research that is defined and maintained by the World Health Organisation WHO. The ATC and DDD system is an internationally agreed code system that can be used to exchange and compare data on drug use at international, national, or local levels.

This MedMetrics Drug Passport is the ultimate documentation that enables many applications that require privacy, transparency, accuracy, and uniqueness. MedMetrics can help to save people's life by providing alerts in drug interactions. There are several types of drug interactions, the main three types are drug-drug/herbal products, drug-disease, and drug-food/alcohol. Software can help clinicians to detect drug interactions, but many programs have not been updated with the evolving knowledge of these interactions, and do not take into consideration important factors such as patients of different age groups, nutritious levels or ethical groups [10, 11].

The following are how MedMetrics Drug is used under different interaction scenarios:

1. For the drug-drug/herbal products interaction. MedMetrics can help using existing records in comparing the ATC code for drug checking. A risk alert will be given when a MedMetrics Drug Passport is paired with the DDD code and the existing ATC code in the MedMetrics Patient Passport.

Purpose of applications	Case studies	MedMetrics empowerment
Medication management	Prescribing	Patient identification
	Clinician-order entry	Clinician identification
	Medication reconciliation	MedMetrics Drug Passport: Dosage comparison from historical records
	Drug-safety alerts	MedMetrics Drug Passport
Documentation	Structured text entry	N/A
	Dictation	N/A
Patient management	Disease management	MedMetrics Patient Passport
	Appointment and testing reminders	N/A
	Care instructions	MedMetrics Patient and Drug Passports
	Result notification	N/A
	Patient behaviour modification	N/A
Quality improvement	Management of patient transfer and transition	MedMetrics Patient passport
	The Healthcare Effectiveness Data and Information Set (HEDIS)	MedMetrics Patient passport
Administrative tools	Billing	N/A
	Referral management	MedMetrics Patient passport
	Risk stratification	MedMetrics Patient and Drug Passports
Communication	Doctor-patient communication	N/A
	Multispecialty or team communication	N/A
	Patient support	MedMetrics Patient and Drug Passports
	Patient or clinician social networking	N/A
Public health reporting	Notifiable disease reporting	MedMetrics Patient passport
	Biosurveillance	MedMetrics Patient passport
	Pharmacosurveillance	MedMetrics Drug passport
Healthcare Industry	Health insurance	MedMetrics Patient passport
	Medical devices	MedMetrics Patient passport and international standards
	Emerging technologies and algorithms	MedMetrics Patient and Drug passports

Table 1.
Categories of substitutable applications, selected examples and MedMetrics [12].

2. For the Drug-condition interaction, the ATC code of the new subscripted drug will pair with the DDD code and the ICD code in the MedMetrics Patient Passport and provide a risk alert, respectively.

3. For people who have an allergy to a specific medication, undertaking drugs with a MedMetrics Drug Passport means their risk of allergy will be reviewed before subscription. Many medicines are branded in different names in the pharmaceutical industry while having similar chemistry comments.

In conclusion, this innovative MedMetrics Drug Passport can help save people's lives, especially for the elderly or the cohort under medication treatment.

2.4 MedMetrics in the digital health era

In 2009, the Journal of New England of Medicine had published a Perspective paper on the health information economy [12]. In this sub-section, I have listed the categories of substitutable applications with selected examples—these examples were chosen to demonstrate how MedMetrics performs in the laboratory, clinical practice, hospital, and home-monitoring environment. Combining these exemplars can form different applications that fit a majority of needs in medical and healthcare environments (**Table 1**).

Another main area that MedMetrics can empower is medical science, pharma industry, and scientific biology research. Medicine is increasingly becoming an interdisciplinary area of medical science, surgical intervention, and an information industry identified by governments, healthcare service providers, health product industry, and patients [13]. In **Table 2**, the author summarised how MedMetrics can contribute to medical science in research, safety, and information transfer.

As shown in **Tables 1** and **2**, among the purpose of applications, MedMetrics plays a critical role in authorisation, security, and policymaking.

Purpose of applications	Case studies	MedMetrics empowerment
Medical Research	Clinical trial eligibility	MedMetrics Patient passport and international standard
	Cohort study tools	International standards
	Electronic data capture for trials	International standards
	Laboratory-test interpretation	International standards
	Genomics	International standards
	Guideline management	International standards
Data acquisition	Laboratory data feed	
	Dispensed medication feed	Blockchain, MedMetrics Patient and Drug passport
	Personally controlled health record data feed	Blockchain and MedMetrics Patient
	Public health data feeds (e.g., local context for infectious diseases)	Blockchain, MedMetrics Patient and Drug passport

Table 2.
MedMetrics in medical science: Research, safety and information transfer [12].

3. MedMetrics, artificial intelligence, blockchain and beyond

The data structure in the healthcare system is highly correlated and complex. Many vendors provide healthcare solutions that are IT-centered instead of patient-centered. Thanks to the nature of the data structure in the MedMetrics Passports for patients and drugs, AI, Cloud service, Internet of Things (IoT), and blockchain methods will be able to play a significant role in the upcoming digital health revolution. In this section, the author will introduce how MedMetrics will work in EHR, fuse with AI, blockchain, and IoT (e.g., wearable sensors).

3.1 MedMetrics in electronic health records

Biometric technologies can offer fast and multiple ways of authentication. The encoding and decoding technology have to apply to information that is generated based on the characteristics of objects. Securing electronic health records could become a complex and costly activity, especially in a scenario where multiple actors potentially maintain information [14, 15].

Remote patient monitoring applications mainly focus on connecting clinicians and patients in hospital, community, and home environments. The ultimate goal is to empower both patients and clinicians with timely information for making necessary interventions at an optimised point of service—it will improve the clinical outcome and reduce the risk of burden on the health economics in the long term.

The European Committee for Standardisation has released a set of information security standards to provide a framework for secures to rage and release of health data [16]. The European standards recognise four global security needs that any health information system should accomplish. They are availability, confidentiality, integrity, and accountability [16].

1. Availability is the main barrier between different hospitals and clinics. The MedMetrics Patient Health Passport, blockchain technology will allow health care providers and governing bodies to re-visit the barrier in patient information availability and consider it an integrated solution.

2. Confidentiality of patient information is a major concern for patient EHR storage and sharing. One way to deal with it is that users with the right to access patient information need to be authorised and allowed to do so in order to perform their duties. In this case, the principle of need-to-know is the key concept to be applied [17]. The other way is to unlock the information that is required for a specific purpose, instead of at the same level. In any case, the information accessed should be relevant but also sufficient to provide health care services [18]. Having an encrypted MedMetrics code can dilute the concern of fraud but cannot entirely remove the concern of restricting the users.

3. Integrity is a blade with both sides. A correctly integrated EHR will help clinicians understand patients' health history with a higher likelihood of making correct clinical interventions. However, a merged patient EHR with a fragment of mixed correct and incorrect will bring uncertainty untold and unseen until the data transformation is done. This can be due to human error from the raw record, but the most challenging part is combining complex historical data among different IT platforms and data structures. The MedMetrics deployed

international standards and codes, removing the integrity concern from its root. However, transferring historical data into MedMetrics is yet challenging.

4. Accountability in ethics and governance is equated with answerability, blameworthiness, liability, and the expectation of account-giving [19]. Accountability is central to the discussion in governance for services in public sectors, nonprofit and private, and individual contexts [20, 21]. The MedMetrics passports for patients and drugs are a secure and comprehensive solution to help deal with the privacy and trustfulness issues in the EHR.

The questions of need-to-know, who-to-know, when-to-know, where-to-know are all driven by the relevance of the acceded information. Defining the correct balance between security requirements and the availability of information is a critical goal in a complex healthcare environment [22]. Relevancy is an ambiguous concept that depends on the context. Having a Medimetrics and biometrics authentication of users makes it possible to guarantee that information is accessed, added, and un-modified by the authorised party.

The figure below demonstrates the principle of MedMetrics and how it might work in the healthcare system (**Figure 2**).

3.2 MedMetrics and artificial intelligence

This sub-section will introduce how AI and blockchain can seamlessly work with the MedMetrics infrastructure. Clinical decision-making support uses AI algorithms to support clinicians, healthcare providers, and insurance companies to make real-time clinical or operational decisions. Some companies have implanted AI in the healthcare pipeline. For example, Sensyne Health is the UK's first public listed company in clinical AI, partnered with Oxford University Hospitals and the University of Oxford. It provides AI solutions to both life sciences challenges and healthcare solutions. Another example is Nuance, a company that partnered with Microsoft to provide Healthcare AI solutions and services.

3.2.1 Natural language processing models

MedMetrics is a group of healthcare codes that reflect the uniqueness of patient and drug information. The data structure means it is suitable for natural language processing (NLP) models such as the Bert model [23], a pre-training of deep

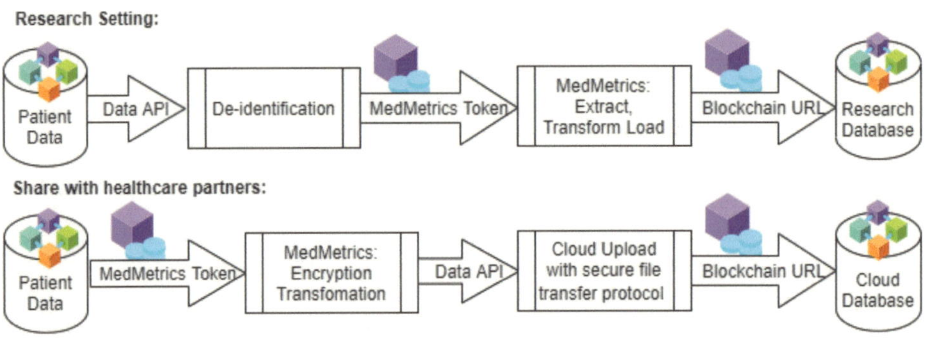

Figure 2.
MedMetrics data pipeline in healthcare.

bidirectional transformers for language understanding. Several Bert models were designed for ICD code. For example, Med-Bert is a method that uses pre-trained contextualised embedding on large-scale structured electronic health records for disease prediction [24]. BEHRT is a deep neural sequence transduction model for EHR that simultaneously predicts the likelihood of 301 conditions in one's future visits [25]. Med-BERT used ICD-9 and ICD-10 codes for diagnosis and the BEHRT used the Clinical Practice Research Datalink (CPRD). The CPRD contains longitudinal primary care covering 35 million patients in the UK. Both Med-BERT and BEHRT proved to be a powerful tools in transferring reactive treatment into preventive medicine at the national level.

3.2.2 Federate learning models

Federated learning is a learning paradigm seeking to address the problem of data governance and privacy by training algorithms collaboratively without exchanging the data itself [26]. Given that scores of data are widely spread across hospitals/individuals, a decentralised computationally scalable methodology is needed [27].

The combination of MedMetrics and Federate Learning will benefit disease prediction studies with small local training datasets, reduce data collection expenses, and accelerate the pace of artificial intelligence-aided healthcare. It will help deal with multi-arm clinical studies across different counties (hence have different data privacy policies) and develop predictive models that require data feeding from time to time.

3.2.3 AI in physiological sensor data

Body Sensor Network (BSN) [28] is one of the most exciting concepts in patient home monitoring. It builds a continuous information hub that can provide real-time and long-distance monitoring. The usage of BSN is still a relatively new area in biometrics. There is a great potential for using one or multiple BSN for personnel identification in healthcare settings.

One of the most mature areas in physiological sensor data biometrics is ECG biometrics. ECG has emerged as a biometrics method with promising results. But same as other sensor technologies, the measurement signals come with sensor failures and measurement variance. On an individual basis, ECG signals can be influenced by physical and psychological changes, such as emotional and mental states, exercise, body position, diets, physical diseases, and positions of electrodes [5]. In addition, signal-acquisition devices produce noises such as from power line interference, baseline wandering, and electrode motion artefacts [29]. Therefore, future work in ECG biometric algorithms is still needed to deal with the concurrence of noises and sample variation.

In applications using ECG as biometrics measures, technologies such as convolutional neural network (CNN), recurrent neural network (RNN), graph regularised non-negative matrix factorisation and sparse representation, One-Dimensional Multi-Resolution Local Binary Patterns, and multi-feature collective non-negative matrix factorisation have been used for pattern recognition (**Figure 3**) [30–35].

3.3 MedMetrics and blockchain

Telemedicine, telehealth, EHR systems, automated retrieval, or update of the electronically stored patient data are common issues that restrict data sharing in the medical science and healthcare sector. Blockchain is a technology in data encryption/

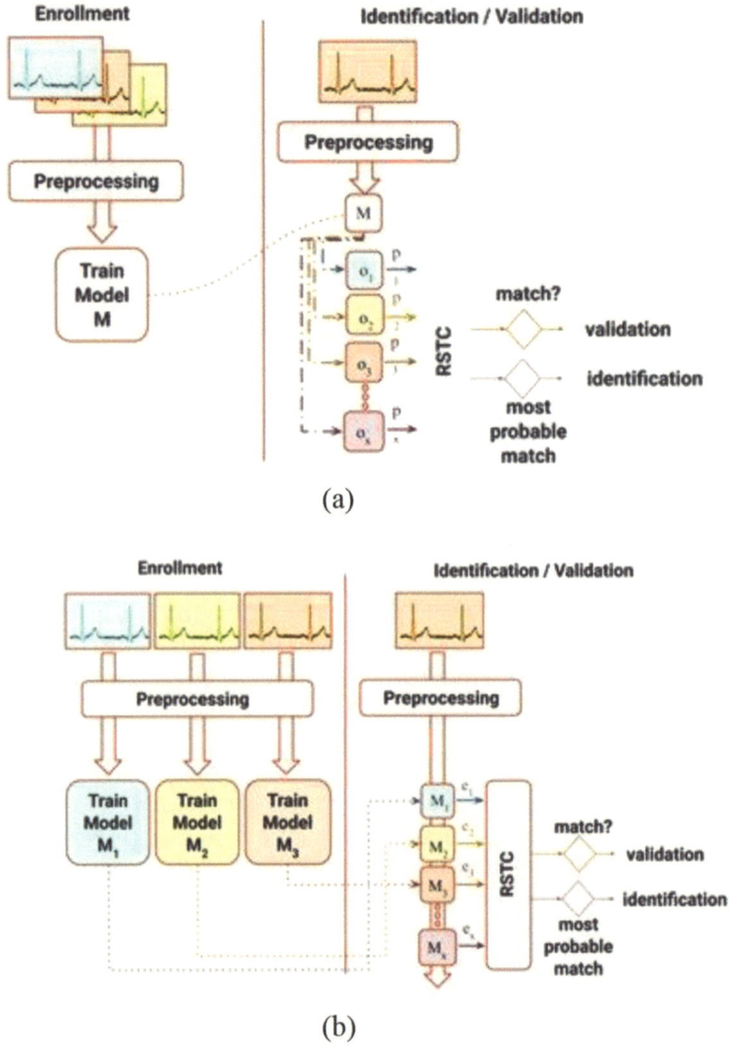

Figure 3.
Examples of deep learning infrastructure for ECG authorisation. (a) an example of a CNN infrastructure, (b) an example of an RNN infrastructure [35].

decryption and tracking. It accelerates innovation, enhances trust, and improves efficiency by overcoming data openness, transparency, and trust concerns. Research suggested that the blockchain might transform how decisions and the interactions between clinicians and patients are recorded [36]. Digital medicine is a field concerned with the use of technologies as tools for measurement and intervention in the service of human health [37]. In the digital medicine environment, blockchain technology will provide AI-based algorithms and applications with the guardian of privacy and authenticity [38].

Due to the size of ever-growing patient data and the future-proof institutional and nationwide service infrastructure, technology developers and health service providers gradually moved towards cloud service. The advancement in cloud networks has enabled connectivity of both traditional networked elements and new devices from IoT [39]. The governance and standard in supporting Cloud service are ready in taking on new healthcare applications. For the Cloud Service, Health Level Seven International (HL7) framework is a globally accepted standard. HL7 is the global authority on standards for interoperability of health technology with members in over 55 countries [40]. Moreover, the Fast Healthcare Interoperability Resources (FHIR) is a standards framework created by HL7 that combines HL7, CDA product lines and web standards.

By using MedMetrics and blockchain, the internationally identifiable MedMetrics code will provide a seamless solution to make patient data transferable, traceable, and interoperable between hospitals and different countries.

4. Conclusion

Clinical science and medical research in healthcare is an evidence-based practice. The three main pillars for the next generation of digital health and the medical information revolution are securely encrypted data (e.g. using blockchain), artificial intelligence, and governance standard and regulations. The information governance, patient health and safety, ethics and fairness, and economic cost all suggest that a more internationally agreed and unified measure is required to support the upcoming digital health revolution.

Biometrics technologies have been broadly used in the public sectors, including the army, door entry, and data access systems, in authorising personnel access and data access for employees. However, in the healthcare sector, biometrics technologies are mainly used by doctors to access patient information (using user names and fingerprints for one-to-one authorisation). There is an emerging trend of using face recognition to log into online health services, but it is more to combine with mobile authorisation, such as Apple face recognition plug-in, instead of healthcare provider.

The slow movement of using biometrics in the health and care sectors is mainly due to the lack of regulation and policy support for patient data protection. There are hundreds of applications available in the commercial market to suggest they can provide healthcare biometrics solutions in hospitals, doctor's offices, and patient's mobile devices for patient data access control. However, the path of transferring patient EHR among devices, platforms, and hospitals is unclear at the law and regulation level. Due to this reason, most biometrics applications are based on individual hospitals, certain medical procedures, or specific clinics. The legal developments in healthcare have been driven by the public concern for personal privacy and confidentiality.

The MedMetrics passports for patients and drugs will help enable and scale up the innovation in the digital health industry. The MedMetrics Drug Passport can avoid preventable medical procedure mistakes, such as giving the wrong medication. By combining with the blockchain, the MedMetrics Patient Passport can provide a secure healthcare data storage and transfer infrastructure.

Beyond the novel solution in dealing with data privacy, the MedMetrics Patient and Drug Passports will help prevent fraud. The technical challenge of these paradigm shifts is interoperability for supporting the delivery of care at multiple locations by multiple carers who need to share the patient health record [41]. By applying AI algorithms, MedMetrics can identify patients with different health conditions as recorded in the previous record. This will help to prevent fraud in health insurance. By combing an AI layer, the MedMetrics can also alert clinicians when the newly collected MedMetrics information is largely different from the previous record, which means that patient may need a medical review. The concept of MedMetrics and its deployment will change the user demographic of healthcare applications that healthcare providers would fuse into their clinical practice based on the current healthcare regulations. The developments in standardisation within digital health will help lower long-term costs in public health and improve the quality of healthcare.

In conclusion, biometrics in medical science is an emerging area. The benefits of using biometric identity verification for healthcare are increasing thanks to emerging technologies in blockchain, IoT, fast-speed mobile networks, and more and more powerful mobile phone devices. However, safely implementing biometrics technologies in applications, especially in healthcare and medical settings, is still more prominent in response to a government-led, regulation-based infrastructure. The concept of MedMetrics can change the user demographic, shift safety restrictions, define new user applications, and encourage ethical AI regulations in medical science and health care. It will boost the next generation of biometrics in medical science and health care and encourage hospital-centered, patient-centered, or service-catered applications.

Acknowledgements

The author thanks Prof. David Clifton for mentoring and supporting her in the Daphne Jackson and Royal Academy of Engineering Career re-entry Fellowship. The author thanks Prof. Chris Chatwin, Dr. Rupert Young, Department of Engineering and Design, University of Sussex, United Kingdom, for their supervision in my D.Phil study on mobile iris biometrics.

This work was supported by the Royal Academy of Engineering, Daphne Jackson Trust, Oxford John Fell Fund (0011028), Wellcome Trust (217650/Z/19/Z) for their funding support.

Conflict of interest

The author declares no conflict of interest.

Author details

Huiqi Yvonne Lu
Department of Engineering Science, Institute of Biomedical Engineering, University of Oxford, United Kingdom

*Address all correspondence to: yvonne.lu@eng.ox.ac.uk

References

[1] Lu H, Chatwin CR, Young RC. Iris recognition on low computational power mobile devices. In: Biometrics-Unique and Diverse Applications in Nature, Science, and Technology. London: InTechOpen; 2011. p. 2

[2] Bud A. Facing the future: The impact of apple FaceID. Biometric Technology Today. 2018;**2018**(1):5-7

[3] Available from: https://www.iproov.com/what-we-do/industries/healthcare [Accessed: 06 Jan 2022]

[4] Odinaka I et al. ECG biometrics: A robust short-time frequency analysis. In: 2010 IEEE International Workshop on Information Forensics and Security. London: IEEE; 2010. pp. 1-6

[5] Pinto JR, Cardoso JS, Lourenço A. Evolution, current challenges, and future possibilities in ECG biometrics. IEEE Access. 2018;**6**:34746-34776

[6] Pouryayevali S, Wahabi S, Hari S, Hatzinakos D. On establishing evaluation standards for ECG biometrics. In: 2014 IEEE International Conference on Acoustics, Speech and Signal Processing (ICASSP). London: IEEE; 2014. pp. 3774-3778

[7] Ingale M, Cordeiro R, Thentu S, Park Y, Karimian N. ECG biometric authentication: A comparative analysis. IEEE Access. 2020;**8**:117853-117866. DOI: 10.1109/ACCESS.2020.3004464

[8] ISO 14155:2020. Clinical Investigation of Medical Devices for Human Subjects—Good Clinical Practice. Geneva, Switzerland: IOF Standardization, ISO; 2020

[9] HEDIS and Performance Measurement', National Committee for Quality Assurance. Available from: https://www.ncqa.org/hedis/ [Accessed: 06 Jan 2022]

[10] Mallet L, Spinewine A, Huang A. The challenge of managing drug interactions in elderly people. The Lancet. 2007; **370**(9582):185-191. DOI: 10.1016/S0140-6736(07)61092-7

[11] Hitchings AW. Monitoring drug therapy. Medicine. 2016;**44**(7):427-432. DOI: 10.1016/j.mpmed.2016.04.004

[12] Mandl KD, Kohane IS. No small change for the health information economy. The New England Journal of Medicine. 2009;**360**(13):1278-1281. DOI: 10.1056/nejmp0900411

[13] I. National Research Council Committee on Engaging the Computer Science Research Community in Health Care. The National Academies Collection: Reports funded by National Institutes of Health. In: Stead WW, Lin HS, editors. Computational Technology for Effective Health Care: Immediate Steps and Strategic Directions. Washington (DC): National Academies Press (US); 2009 Copyright ©, National Academy of Sciences, 2009

[14] Agrawal R, Johnson C. Securing electronic health records without impeding the flow of information. International Journal of Medical Informatics. 2007;**76**(5-6):471-479

[15] Flores Zuniga AE, Win KT, Susilo W. Biometrics for electronic health records. Journal of Medical Systems. 2010;**34**(5):975-983. DOI: 10.1007/s10916-009-9313-6

[16] Klein GO. Health informatics—Security for healthcare communication,

E. C. F. Standardization. Methods of Information in Medicine. 2002;**41**(4): 261-270

[17] Blobel B. Authorisation and access control for electronic health record systems. International Journal of Medical Informatics. 2004;**73**(3):251-257

[18] van der Linden H, Kalra D, Hasman A, Talmon J. Inter-organizational future proof EHR systems: A review of the security and privacy related issues. International Journal of Medical Informatics. 2009;**78**(3):141-160

[19] Dykstra CA. The quest for responsibility. American Political Science Review. 1939;**33**(1):1-25

[20] Shahib HM. Towards Local Government's Integrated Accountability Framework. In: Towards the Local Government's Integrated Accountability Framework. Berlin/Heidelberg, Germany: Springer; 2021. pp. 115-131

[21] Srivastava V, Mahara T, Yadav P. An analysis of the ethical challenges of blockchain-enabled E-healthcare applications in 6G networks. International Journal of Cognitive Computing in Engineering. 2021;**2**:171-179

[22] Blobel B. Application of the component paradigm for analysis and design of advanced health system architectures. International Journal of Medical Informatics. 2000;**60**(3):281-301

[23] Devlin J, Chang M-W, Lee K, Toutanova K. Bert: Pre-training of deep bidirectional transformers for language understanding. In: Proceedings of NAACL-HLT. 2019. pp. 4171-4186

[24] Rasmy L, Xiang Y, Xie Z, Tao C, Zhi D. Med-BERT: pretrained contextualized embeddings on large-scale structured electronic health records for disease prediction. NPJ Digital Medicine. 2021;**4**(1):1-13. DOI: 10.1038/s41746-021-00455-y

[25] Li Y et al. BEHRT: Transformer for electronic health records. Scientific Reports. 2020;**10**(1):1-12

[26] Rieke N et al. The future of digital health with federated learning. NPJ Digital Medicine. 2020;**3**(1):1-7. DOI: 10.1038/s41746-020-00323-1

[27] Brisimi TS, Chen R, Mela T, Olshevsky A, Paschalidis IC, Shi W. Federated learning of predictive models from federated electronic health records. International Journal of Medical Informatics. 2018;**112**:59-67. DOI: 10.1016/j.ijmedinf.2018.01.007

[28] Lo BP, Thiemjarus S, King R, Yang G-Z. Body Sensor Network—A Wireless Sensor Platform for Pervasive Healthcare Monitoring. London: IEEE; 2005

[29] Limaye H, Deshmukh V. ECG noise sources and various noise removal techniques: A survey. International Journal of Application or Innovation in Engineering & Management. 2016;**5**(2): 86-92

[30] Donida Labati R, Muñoz E, Piuri V, Sassi R, Scotti F. Deep-ECG: Convolutional Neural Networks for ECG biometric recognition, Pattern Recognition Letters. Vol. 126. 2019. pp. 78-85. DOI: 10.1016/j.patrec.2018.03.028

[31] Louis W, Komeili M, Hatzinakos D. Continuous authentication using one-dimensional multi-resolution local binary patterns (1DMRLBP) in ECG biometrics. IEEE Transactions on Information Forensics and Security. 2016;**11**(12):2818-2832. DOI: 10.1109/TIFS.2016.2599270

[32] Li R, Yang G, Wang K, Huang Y, Yuan F, Yin Y. Robust ECG biometrics using GNMF and sparse representation.

Pattern Recognition Letters. 2020;**129**:70-76. DOI: 10.1016/j.patrec.2019.11.005

[33] Ibtehaz N et al. EDITH: ECG biometrics aided by deep learning for reliable individual authentication. IEEE Transactions on Emerging Topics in Computational Intelligence. 2021;**1**:1-13. DOI: 10.1109/TETCI.2021.3131374

[34] Huang Y, Yang G, Wang K, Liu H, Yin Y. Robust multi-feature collective non-negative matrix factorization for ECG biometrics. Pattern Recognition. 2022;**123**:108376. DOI: 10.1016/j.patcog.2021.108376

[35] Belo D, Bento N, Silva H, Fred A, Gamboa H. ECG biometrics using deep learning and relative score threshold classification. Sensors. 2020;**20**(15):4078. Available from: https://www.mdpi.com/1424-8220/20/15/4078

[36] Leeming G, Ainsworth J, Clifton DA. Blockchain in health care: Hype, trust, and digital health. The Lancet. 2019;**393**(10190):2476-2477

[37] Goldsack JC et al. Verification, analytical validation, and clinical validation (V3): the foundation of determining fit-for-purpose for Biometric Monitoring Technologies (BioMeTs). npj Digital Medicine. 2020;**3**(1):55. DOI: 10.1038/s41746-020-0260-4

[38] Elenko E, Underwood L, Zohar D. Defining digital medicine. Nature Biotechnology. 2015;**33**(5):456-461

[39] Faizullah S, Khan MA, Alzahrani A, Khan I. Permissioned Blockchain-Based Security for SDN in IoT Cloud Networks. In: 2019 International Conference on Advances in the Emerging Computing Technologies (AECT). London: IEEE; 2020. pp. 1-6. DOI: 10.1109/AECT47998.2020.9194181

[40] Elena Vega D. Automated interoperability testing of healthcare information systems. In: Memon A, editor. Advances in Computers. Vol. 85. Amsterdam, Netherlands: Elsevier; 2012. pp. 213-276

[41] Shoniregun CA, Dube K, Mtenzi F. Laws and standards for secure e-healthcare information. In: Electronic Healthcare Information Security. Berlin, Germany: Springer; 2010. pp. 59-100

Chapter 7

Quantum Biometrics

Iannis Kominis, Michail Loulakis and Özgur E. Müstecaplıoğlu

Abstract

It was recently proposed to use the human visual system's ability to perform efficient photon counting in order to devise a new biometric authentication methodology. The relevant "fingerprint" is represented by the optical losses light suffers along different paths from the cornea to the retina. The "fingerprint" is accessed by interrogating a subject on perceiving or not weak light flashes, containing few tens of photons, thus probing the subject's visual system at the threshold of perception, at which regime optical losses play a significant role. The name "quantum biometrics" derives from the fact that the photon statistics of the illuminating light, as well as the quantum efficiency at the light detection level of rod cells, are central to the method. Here we elaborate further on this methodology, addressing several aspects like aging effects of the method's "fingerprint," as well as its inter-subject variability. We then review recent progress towards the experimental realization of the method. Finally, we summarize a recent proposal to use quantum light sources, in particular a single photon source, in order to enhance the performance of the authentication process. This further corroborates the "quantum" character of the methodology and alludes to the emerging field of quantum vision.

Keywords: quantum, biometrics, photon statistics, quantum light, visual perception

1. Introduction

It was recently proposed [1] to use the human visual system's ability to perform photon counting in order to devise a new biometric authentication scheme, which was called "quantum." The claim made in [1] was that the scheme offers unbreakable security, not unlike the security offered by quantum cryptography [2, 3] against a potential impostor wishing to eavesdrop during the transmission of information. In our case, the "fingerprint" is a physical property of the visual system, including the eyeball, retina and brain. The "fingerprint" is registered and probed using weak-intensity light and the subject's conscious perception thereof.

In this chapter we will further elaborate in intuitive terms on the workings of the quantum biometric methodology as were outlined in [1]. To do so, we will summarize a recently proposed authentication algorithm [4], which is straightforward to understand, as compared to more elaborate algorithms discussed in [1]. We will then address some basic issues of the authentication methodology. One has to do with the very first registration of one's "fingerprint." Another issue is related to aging effects on this "fingerprint," which have to do with the visual acuity degrading with age.

We will also address the central issue of the variability of the "fingerprint" among different individuals.

We will then review recent progress made towards the experimental realization of the quantum biometric methodology using laser light [5]. Finally, we will summarize a recent proposal [4], to use quantum light in order to enhance the method's performance in terms of the required time to run the authentication algorithm, for given desired values of the false-negative and false-positive authentication probability.

2. Preliminaries

As a short introduction to the basics of our biometric authentication methodology, we first recapitulate the original experiment of Hecht et al. [6], eloquently described by Bialek [7]. In particular, Hecht et al. were the first to unambiguously demonstrate that rod cells, the scotopic photoreceptors in our retina, are efficient photon detectors. Additionally, they obtained the threshold in the number of detected photons for the perception of vision to take place. We denote this threshold by K, and from the work of [6] it follows that $K \approx 6$. We note that a recent psychophysical experiment performed by Vaziri and coworkers [8] using a single photon source for the stimulus light found that $K \approx 1$. We will here use the previously accepted value of $K = 6$, and defer to future work the analysis of our methodology's dependence on the precise vale of K, which still is a rather complex open problem.

In more detail, the three authors in [6] exposed their eyes to very weak-intensity light pulses, with the photon number within each pulse being so small, that the visual perception became a probabilistic event. Let P_{see} be the probability of seeing such a light pulse. An expression for P_{see} can be found as follows. Denote by $\tilde{N}$ the mean number of photons within a light pulse of duration τ and intensity I, that is, $\tilde{N} = I\tau$. We know that coherent light has Poissonian photon statistics, that is, the probability to have exactly n photons within such a pulse is $\tilde{N}^n e^{-\tilde{N}}/n!$.

However, when the mean number of photons per pulse incident on the eyeball is $\tilde{N}$, the mean number of photons per pulse being incident on the retina is smaller by a factor α_l, where $0 < \alpha_l < 1$. This factor describes the optical losses suffered by light along its path from the cornea to the retina. Moreover, from those photons incident on the retina, only a fraction α_d will be detected by the illuminated rod cells, one reason being the finite quantum efficiency of the rod photoreceptors. All those factors can be lumped together in a single factor, call it α, quantifying the various sources of optical loss. Then, the mean number per pulse of *actually detected* photons will be $\alpha\tilde{N}$, where $\alpha = \alpha_l \alpha_d$.

Hence the probability that the number of photons detected by the illuminated patch of the retina is exactly n is given by $\left(\alpha\tilde{N}\right)^n e^{-\alpha\tilde{N}}/n!$. If this number is larger than the detection threshold K, the perception of "seeing" a spot of light will take place. Hence,

$$P_{\text{see}} = \sum_{n=K}^{\infty} \left(\alpha\tilde{N}\right)^n e^{-\alpha\tilde{N}} n! \tag{1}$$

As noted by Bialek [7], this formula expresses the (perhaps surprising) fact that the probabilistic nature of our visual perception, which is a systemic effect concerning the retina and the brain, is fundamentally governed by the quantum statistical properties of the stimulus light.

To further understand the experiment of Hecht et al., we plot in **Figure 1** examples of the dependence of the probability P_{see} on the mean number of photons per pulse incident on the cornea, $\tilde{N}$. In **Figure 1a** we keep the threshold K and vary α, whereas in **Figure 1b** we keep α constant and vary K. Both dependences are rather obvious to interpret. Regarding **Figure 1a**, it is evident that for a given perception threshold K, higher optical loss (small α) requires a higher photon number $\tilde{N}$ for the perception of light to be highly probable. Similarly, regarding **Figure 1b** it is seen that given an optical loss factor α, the smaller the threshold K the fewer photons are needed to obtain a given value of P_{see}.

What is interesting to note is that the change of α (**Figure 1a**) leaves the overall shape of the functional dependence of P_{see} versus $\tilde{N}$ pretty much invariant, that is, it roughly brings about a translation of the figure along the x-axis. In contrast, the change of K qualitatively changes the shape of the dependence of P_{see} versus $\tilde{N}$. Now, what the authors in [6] observed was that although each one of the three authors participating in the measurement produced a different dependence of P_{see} versus $\tilde{N}$, all three curves could be coalesced by such a translation along the x-axis, and all could be fit with a common value of $K \approx 6$. This is shown in Figure 2.2 of [7].

The experimental apparatus used by Hecht et al. looks rather primitive from our modern technological perspective. Yet these authors managed to make a remarkable case: even though a subjective observable, as the optical loss parameter α, which changes among individuals, perplexes the analysis of individuals' responses to perceiving or not faint light pulses, there appear two objective properties of the human visual system: The first has to do with the wiring of the ganglion cells which communicate visual responses to the brain, and which wiring determines the perception threshold K. The second is what Hecht et al. nicely demonstrated: retina's photoreceptors are efficient single photon detectors. This follows from the fact that the experimentally extracted number of detected photons is much smaller than the

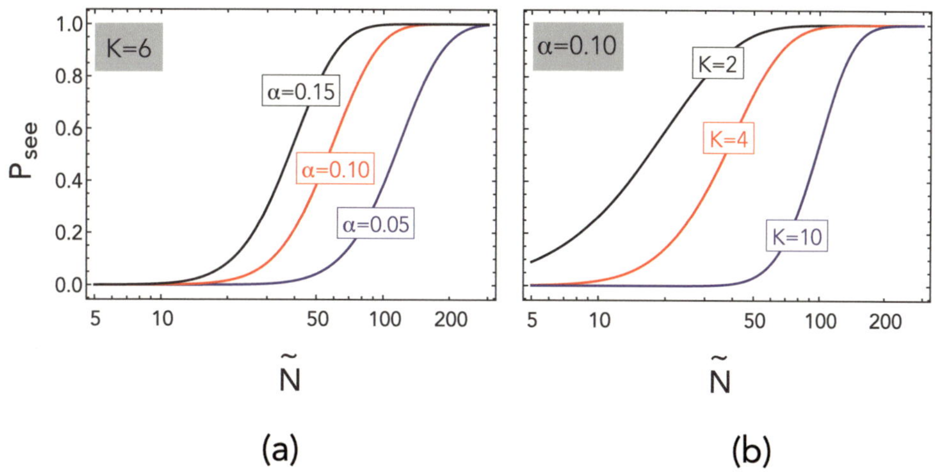

Figure 1.
Probability of seeing a light pulse having mean photon number per pulse $\tilde{N}$ versus $\tilde{N}$, as calculated from Eq. (1), for (a) various values of the optical loss parameter α, and constant perception threshold K, and (b) various values of K and constant α. It is seen that for constant K, a change in α practically translates the curve along the x-axis, while for a given α, a change in K alters the shape of the curve.

number of illuminated rod cells. It took several years until the photo-detection properties of rod cells were unraveled with modern quantum-optical techniques [9–15].

3. Quantum biometrics

Whereas the variability of the parameter α among individuals was a nuisance for Hecht et al., who wanted to demonstrate the single-photon-detection capability of rod cells, we now take this physiological capability for granted, and instead use the variability of α as a biometric quantifier. However, a single number cannot offer a useful biometric "fingerprint." Hence the idea analyzed in [1] was to use a whole map of α values, the so-called α-map, by considering several different paths of light towards the retina. Here we elaborate a bit more on this.

There are three ways to get several light paths to the retina. For all three we suppose that the stimulus light source consists of distinct laser beams, which can illuminate the cornea at several different spots (as shown in **Figure 2a**), either one at a time, or many. These laser beams are supposed to propagate in parallel from the light source to the cornea. Then, for an emmetropic individual (i.e., somebody not having any refractive errors) all these laser beams will be focused on the same spot on the retina. Instead, for a myopic individual these laser beams focus before the retina and thus will illuminate different spots on the retina, while for a hyperopic individual they focus behind the retina, and again illuminate different spots.

Now, as observed in Section 2, the α factor quantifies both the optical losses, α_l, suffered by light along its path from the cornea to the retina, and the probability of photon detection, α_d, once the photons reach the retina. Thus, for an emmetropic subject the difference in α between different laser beams stems only from the difference in α_l, while for myopic or hyperopic individuals the difference in α stems from both the different α_l and the different α_d for each laser beam. For the authentication algorithm to work, we need the perception of different patterns of

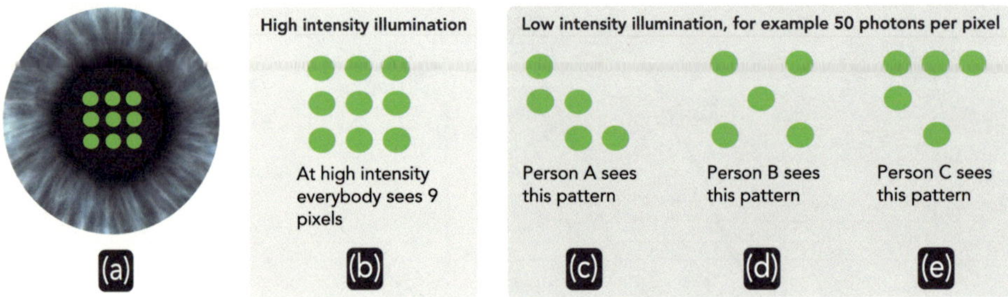

Figure 2.
A simplified presentation of the idea behind the biometric authentication using the photon counting capability of the human visual system. (a) We suppose to have a light stimulus source, which can provide for parallel laser beams patterned in an array. For simplicity, this is here shown as a 3×3 array. The laser beams propagate in parallel from the source to the eye, being incident on the cornea. We further suppose that all of them, as shown in (a), or a subset of them can be simultaneously illuminated during a given pulse. If one could see the reflection of the laser beams off the cornea, one would see the image (a), where the iris is shown to be illuminated by nine spots. (b) If the illuminated laser beams contain a large number of photons per pulse, and further assuming that all human subjects being illuminated are myopic, then all of them would report seeing nine different spots patterned in such a 3×3 array. (c-e) However, if the number of photons per beam per pulse is small, for example in the regime of 5–200 photons, then the visual perception would be working close to its threshold. In that case, the optical losses suffered by light along these nine different paths, different among path-to-path for each individual, and different for a "geometrically similar" path among individuals, will result in a different pattern of spots being perceived by each subject.

simultaneously illuminated spots on the retina. Therefore, and for having a common presentation in the following, we assume that our laser beams are focused on different spots on the retina. This can be optically designed to be the case even for emmetropic subjects.

In **Figure 2** we show the crux of the matter: suppose we have an array of, for example, nine laser beams, patterned in a 3×3 matrix (**Figure 2a**). Further suppose that these beams are all illuminated simultaneously for a given time interval (what we previously referred to as laser pulse), and moreover, let us assume that the mean photon number per beam per pulse is very large, say $\gg 100$ photons. In such a scenario *every* subject (without any visual deficiency) will report seeing nine spots (**Figure 2b**). This is because with certainty, everybody will perceive a pulse containing a large number of photons (far right in the curves of **Figure 1**, where $P_{\text{see}} \approx 1$). However, as we reduce the mean photon number per laser beam per pulse, and move to the regime of the visual threshold described by the variable P_{see} in **Figure 1**, each individual will report different patterns of perception, as shown in **Figure 2b–d**. This difference in perception in the regime of the visual threshold is exactly what our biometric authentication scheme takes advantage of.

To describe the workings of the methodology in more detail, we first note that the prerequisite is that the α-map of the subject that will need to be authenticated by the biometric device has been already measured and stored. This is like taking a subject's fingerprint and registering it in the relevant database. Now, for our biometric methodology this step is the most time consuming step, because the α values for several different light paths must be measured. However, apart from aging effects to be discussed in the following, this step is required only once.

3.1 First registration of α-map

The α-value of a retinal spot can be estimated indirectly through Eq. (1) by measuring the percentage of times light is perceived when the spot is repeatedly illuminated. Precisely, suppose that a spot is illuminated $M \gg 1$ times with coherent light pulses of mean photon number $\tilde{N}$ and that light is perceived in m of these times. The fraction m/M is an experimental proxy for P_{see}, and α can be estimated by solving the equation

$$\frac{m}{M} = \sum_{n=K}^{\infty} \frac{\left(\alpha\tilde{N}\right)^n e^{-\alpha\tilde{N}}}{n!}.$$

To avoid amplifying the error made in the estimation of P_{see} by m/M, one should choose $\tilde{N}$ so that the slope of the right hand side of Eq. (1) is maximal. This is achieved when $\alpha\tilde{N} = K - 1$. Clearly, we cannot choose $\tilde{N}$ based on this condition since α is unknown. Nevertheless, this condition is equivalent to $P_{\text{see}} = \sum_{n=K}^{\infty} \frac{(K-1)^n e^{-(K-1)}}{n!}$. For $K = 6$, this gives $P_{\text{see}} \simeq 0.384$. Hence, as a rule of thumb, a good practice to estimate the α-value of a spot is to use such a laser intensity that light is perceived roughly 40% of the time.

Let us denote by $\hat{\alpha}$ the estimate of α derived in this way. It turns out that an approximate 0.99-confidence interval for α would be $\hat{\alpha} \pm 1.428\hat{\alpha}/\sqrt{M}$. That is, to determine α with 99% confidence, we would roughly need $M = 200$ pulses for a 10% error tolerance, and $M = 800$ pulses for a 5% error tolerance.

This number of interrogations is clearly impractical. In [1] the first authentication algorithm proposed follows a similar route of estimating α, making the process even more impractical, since several such time-consuming series of many interrogations would be needed, one for the registration, and one each time the subject needs to be authenticated.

This observation had motivated [1, 4] authentication algorithms that, rather than using the precise α-values of retinal spots, only require that the α-value of the illuminated spots be above the threshold $\alpha_{\rm hi}$, or below the threshold $\alpha_{\rm lo}$. For such algorithms, one only needs to construct a coarse α-map, in which retinal spots are classified into high-α ($\alpha > \alpha_{\rm hi}$), low-α ($\alpha < \alpha_{\rm lo}$), or intermediate-α ($\alpha_{\rm lo} < \alpha < \alpha_{\rm hi}$) spots. As will be shown in a forthcoming work, a much smaller number of interrogations, typically between 10 and 40, is sufficient to classify a retinal spot. Having done so, that is, knowing the subject's α-map, we can then proceed with elaborating on the authentication process. Before doing so, we make some general comments.

3.2 Detailed description

When the subject wishes to be authenticated, for example, in order to enter a high-security facility, the biometric device must implement a measurement protocol in order to positively authenticate the subject. As already apparent, we have restricted the discussion to authentication. That is, we assume that when asking to be authenticated, the subject announces who he or she is. Then the device must make sure that the subject indeed is who he or she claims to be. So henceforth we suppose the biometric device is "aware" of the subject's α-map.

The result of the authentication protocol is either positive or negative, and two central quantifiers of its performance are the false-negative and false-positive probability, denoted by p_{fn} and p_{fp}, respectively. The former is the probability that a subject truly claiming to be who he or she is, is *not* identified as such. The latter is the probability that an impostor, falsely claiming to be somebody else is positively identified as that other person. Obviously, the more time is taken by the authentication process, the smaller these two probabilities should become. Hence a third important performance parameter is the time required to achieve a given desired value for p_{fn} and p_{fp}.

Let us call Alice the subject who appears and wishes to be positively authenticated. Eve will be an impostor who maliciously claims to be Alice. Now, the biometric device knows Alice's high-α and low-α spots. Hence if the former are illuminated, Alice is expected to perceive light. In contrast, if the low-α spots are illuminated, Alice is expected not to perceive light.

Now, we will suppose that Eve is not aware of Alice's α-map. Is this a fair assumption? Indeed, we can first rightfully suppose that Eve does not have access to the α-maps stored in the biometric device. If we do not make such an assumption, then pretty much all biometric identification methods are vulnerable to counterfeiting. Can Eve otherwise obtain Alices's α-map? Well, the only way that this seems feasible is by use of force, that is, Eve purchases the biometric device and forcefully has Alice undergo a measurement of her α-map, and moreover Eve has found a way to enforce Alice's truthful answers to Eve's device interrogations on light perception. However, we can again safely assume that use of force is not something any biometric device can cope with. For that matter, even quantum cryptography would be irrelevant as a technology, since if somebody enters Bob's office while Bob is securely transmitting information to Alice via a quantum communication channel, this somebody could forcefully obtain the information Bob wants to transmit, hence the quantum communication channel would obviously be of little help. So it is rightful to assume that the biometric "fingerprint" cannot be stolen from the device where it is stored, and use of force is not considered when comparing the performance of biometric authentication methodologies.

However, what should be allowed as a scenario is for the impostor to have technology that would allow her to estimate the "fingerprint" under consideration by physical means, which do not require access to the fingerprint database nor do they require use of force. For example, one could imagine when discussing, for example, face recognition, that Eve could take an image of Alice's face without Alice noticing (e.g., from a distance using a high resolution camera) and then use this image to construct a face mask. This scenario is not prevented by physical laws. Nor is there any physical law preventing the face recognition test from being bypassed by an artificial face mask. So in comparing the security of various biometric methodologies, one should study what is in principle possible in terms of bypassing the biometric device, given the laws of physics. Based on current quantum technology, it is inconceivable how Eve would be able to infer Alice's α-map by physical means, although some comments where made in [1] along this line of thought.

In other words, it seems that even in principle, that is, based on the laws of physics and in particular the physics of quantum measurements, Eve cannot physically obtain Alice's α-map. This is one main advantage of this biometric methodology. In any case, the only option left to Eve when impersonating Alice is to second guess the biometric device's interrogations. Is this possible? Can Eve know whether the device is illuminating a low-α or a high-α spot of Alice, and thus tune her responses accordingly? The answer is negative. The spots being illuminated are randomly chosen by the device, and as far as Eve is concerned, they could be of any kind.

A crucial detail is that the device illuminates every spot, no matter of what kind it is, with a light pulse *always having the same mean number of photons per pulse*. Thus, even if Eve is equipped with a perfect photon counter while she is taking the test, she would just measure light pulses with a given mean number of photons. This measurement does not convey to her any useful information. Further, since she is not aware of Alice's α-map, even if Eve is equipped with a perfect position-sensitive photon detector, she still cannot extract any useful information from any stimulus light patterns emitted by the biometric device. *Eve is forced to respond randomly to the device's interrogations on whether the subject does perceive or does not perceive the light flashes.*

We will now elucidate all of the above using the specific authentication protocol outlined in [4].

3.3 Authentication protocol

This protocol is a variant, which is intuitively simpler to understand than the protocols discussed in [1]. We assume that the biometric device simultaneously illuminates N different retinal spots, some of which are low-α spots, with the rest being high-α spots. The subject taking the test is then questioned on how many spots she perceived. Let H be the random variable quantifying how many high-α spots where illuminated. Further, let R be a random variable quantifying the number of bright spots perceived by the interrogated subject. We define as correct the response for which $R = H$. As will be shown in the following, a single interrogation is not enough to obtain the desired performance metrics, therefore multiple interrogations will be used.

Now the probability that an impostor called Eve, pretending to be Alice, correctly responds to such an interrogation is

$$P_E = 1/(N+1), \tag{2}$$

because Eve is not aware of what kind of spots are being illuminated, and H can take any value between 0 and N, therefore Eve's chance to guess this number correctly is $1/(N+1)$. In contrast, Alice has a much larger probability to successfully respond. To fail, Alice should perceive a low-α spot, or not perceive a high-α spot, with these two errors not canceling out. It turns out [4] that the probability of Alice's successful response is

$$P_A = \frac{1-(1-u)^{N+1}}{(N+1)u} \tag{3}$$

where $u = p_H + p_L$, with p_H being the probability that Alice fails to perceive a stimulus on a high-α spot, and p_L being the probability that Alice does perceive a stimulus on a low-α spot.

Now, as previously mentioned, one interrogation is not enough to achieve adequate performance with respect to the false-positive and false-negative probabilities. Therefore a number of sequential interrogations is used. This number is actually a random variable, coming about as follows [4]. We define an integer success variable S, initiated to $S = 0$ at the beginning of the authentication process. Then if the subject responds correctly, S is increased by 1, whereas it is decreased by 1 if the subject responds wrongly. Positive authentication is established when S reaches a predefined positive value S_+, whereas negative authentication is established when S reaches a predefined negative value S_-. The value S_+ is determined by the required false-positive probability p_{fp}, and the value S_- by the desired false-negative probability p_{fn}. Thus, the random variable S performs a random walk. If the interrogated subject is indeed Alice, then the probability for a positive step of S is P_A given by Eq. (3), and correspondingly, the probability for a negative step is $1 - P_A$. Similarly, if the interrogated subject is Eve (who claims to be Alice), then the respective probabilities for a positive and a negative step are P_E given by Eq. (2) and $1 - P_E$.

For relatively small values of the parameter u, it is $P_A > 1/2$, and Alice's random walk drifts towards positive S. For a number of illuminated spots $N > 2$ it is $P_E < 1/2$, thefore Eve's random walk drifts towards negative S. The smaller the desired p_{fp}, the larger will be S_+, and the more difficult will be for Eve's success parameter to reach the positive authentication value S_+. Similarly, the smaller the desired false-negative probability, the more negative will be S_-, and the more difficult will be for Alice to fail the test. Incidentally, the highest priority for the interrogation is that an impostor will fail the test, that is, the highest priority is the smallness of p_{fp}. The smallness of p_{fn} is also important, but mostly of practical interest. This is because in the unfortunate circumstance that the true subject, Alice, fails the test, she would have to retake it. This will happen the more infrequently, the smaller is p_{fn}.

3.4 Optimal photon number

The reader might have inquired how the photon number per pulse per illuminated pixel is chosen. This is easily shown by considering the fact that the probability of Alice's successful response, P_A, is higher the smaller the parameter u is. Using the probability-of-seeing expression of Eq. (1), one can calculate u as a function of the incident photon number $\tilde{N}$. It should be clear why there is a minimum in such a dependence. For very large $\tilde{N}$, the probability of seeing tends to 1, therefore Alice will for sure perceive illuminated low-α spots, therefore p_L will tend to 1. Similarly, for too small $\tilde{N}$, Alice will have a hard-time perceiving even the illuminated high-α spots, therefore p_H will tend to 1. In either extremes, u will tend to 1, and it becomes minimum for some intermediate value of $\tilde{N}$, which is about 60–80 photons per pulse [4].

4. Aging effects

One question recurring in presentations of the above scheme is the effect of aging, namely, it is reasonable to assume that the α-map of a subject will change with time, like the visual acuity does. Thus it is expected that the α values will become smaller with the subject's age. Would this affect the authentication scheme? To address this question, we will use data from visual perimetry, in particular, differential threshold perimetry. This is a technique used to measure the sensitivity of one's visual field and the construction of the so-called hill-of-vision. The technique is illustrated in **Figure 3**.

The subject fixates at the center of a half-sphere, the inner surface of which has a light background illumination (**Figure 3a**). Then, several spots are illuminated with varying intensity (on top of the background), and the subject reports whether he or

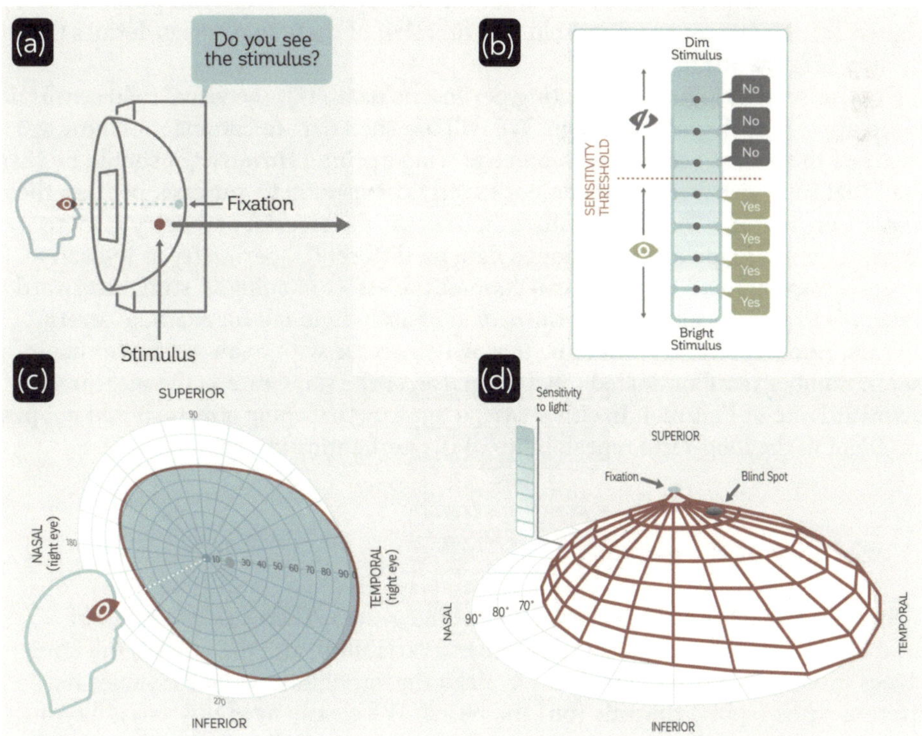

Figure 3.
Measurement of the visual field using differential threshold perimetry. Figure reproduced from [16].

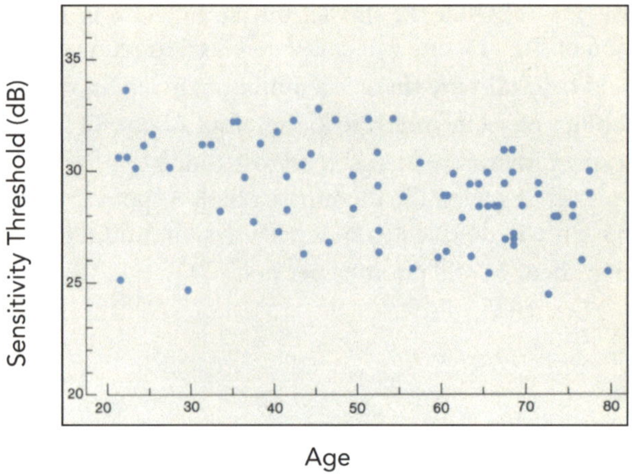

Figure 4.
Sensitivity threshold decreasing with age. Plotted is the average threshold versus age for a particular position (3° nasal, 15° superior) for 74 individuals. Despite the scatter, the downward slope is obvious. Based on such data, it is reasonable to expect that the values of an individual's α-map will similarly decrease with time. This necessitates either a periodic registration of one's α-map, or a gradual increase with one's age of the photon number used per illuminated pixel. Reproduced with permission from American Medical Association [17]. Copyright (1987) American Medical Association. All rights reserved.

she perceives the illuminated spot (**Figure 3b**), this leading to the threshold of perception. The position of each spot is defined with two angles, one accounting for the temporal vs. nasal position, and the other for the superior vs. inferior position (**Figure 3c**). The measured threshold as a function of these two angles defines the hill of vision (**Figure 3d**).

Now, as seen in **Figure 4** depicting perimetric data [17], the visual field sensitivity indeed appears to degrade with age. We will use such data to comment on how age can affect the α-map used as our biometric "fingerprint." However, it should be first noted that such visual-field data do not exactly correspond to our case, because they are not fully scotopic. As the literature on scotopic differential perimetry is more sparse, we will use the aforementioned data on differential perimetry as indicative. In any case, there are two ways one can counter the effect of aging. A straightforward strategy is to periodically register the α-map of an individual, for example, every 10 years. Another strategy would be to slowly increase with one's age the optimal photon number per illuminated pixel per pulse, at the same rate as the measured downward rate of **Figure 4**. In either case, it appears that aging effects should not pose a problem in the long-term repeatability of the authentication process.

5. Variability of the α-map

Another crucial issue is the variability of the α-map. There are two kinds of variability of interest, one is the intra-subject variability, and the other is the inter-subject variability. With the former we mean the variability in one's α values for different paths (spots) towards (on) the retina. We clearly need this variability in order to be able to define in the first place the α-map including high-α, low-α and intermediate-α values. The latter is the variability of the α-map among different subjects, in particular the variability of the α values among individuals for geometrically similar spots on the retina. In **Figure 5** we again show perimetric data [17] accounting for both types of variability.

Figure 5a depicts the variability of the differential threshold of one particular individual for various viewing angles in the central 30° field-of-view. The observed variability from 2 dB up to 6 dB is enough to provide for the definition of an α-map useable for our biometric methodology. **Figure 5b** shows the inter-subject variability, which again ranges from 2 to 6 dB, enough to be able to support our protocols.

Finally, related to the inter-subject variability is the question of how many different subjects would our methodology be able to authenticate without the possibility of a random coincidence of one's α-map with somebody else's. In the next section we will discuss recent experimental progress towards realizing the quantum biometric methodology. There it will be shown that the laser stimulus we developed in [5] provides for a laser beam consisting of a pattern of 5×5 pixels, so 25 pixels in total. Assuming that (i) we classify each pixel with three possibilities, that is, low-α, intermediate-α and high-α, (ii) we use only low-α and high-α values for our authentication protocols, (iii) each three possibilities for the α-values occur with the same probability of 1/3, and (iv) distribution of each kind of α-value is random across the retina, we can estimate the number of possible users of such a biometric device is 10^5. With 50 pixels this number becomes 10^{10}.

6. Spatially selective laser light stimulus

The stimulus light source required to realize an authentication algorithm such as the one described above was recently reported in [5]. It consists of two laser beams,

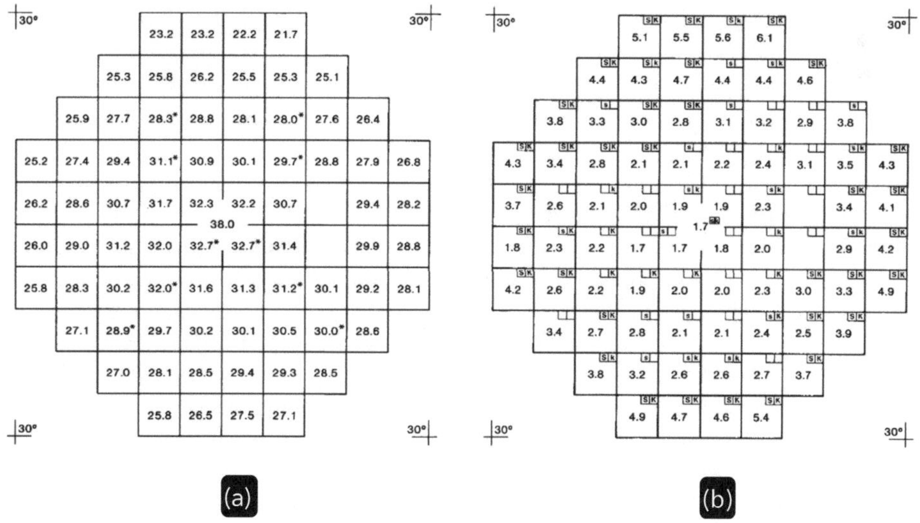

Figure 5.
(a) Intra-subject variability the differential perimetric threshold for the central 30° field-of-view. Numbers reflect the threshold intensity in dB. These data indicate that the variation of the α-values across a single individual's visual field ranges between 2 and 7 dB, which should be enough to define a useable α-map. (b) Inter-subject variability of the differential perimetric threshold for the central 30° field-of-view. Numbers are average inter-subject differences of the threshold intensity in dB. These data the variation among individuals ranges between of 2 and 6 dB, and should be enough to differentiate spatial patterns of weak light perception among different individuals, which differentiation underlies the authentication protocol. Reproduced with permission from American Medical Association [17]. Copyright (1987) American Medical Association. All rights reserved.

one at 532 nm and one at 850 nm, which are combined in a fiber into a single beam. As the laser power at the exit of the fiber combiner fluctuates, in [5] a feedback loop was used to stabilize the power of the 532 nm, which is used as stimulus light. The infrared light is used for pointing, as will be described shortly. In order to create different patterns of pixels across the laser beam's cross section, the laser beam was propagated through a liquid crystal display (LCD) in a multi-pass configuration. The activated dots of the LCD produced an optical loss in the laser beam, corresponding to dark pixels, whereas the inactivated dots produced the illuminated pixels. In order for the contrast between illuminated versus dark pixels to be acceptable, the beam went through the same configuration of LCD dots five times, as shown in **Figure 6a**. The five passes where chosen because the relative optical loss obtained from one pass between activated and inactivated LCD dot is 0.35. Now, since we need photon numbers up to 200 photons per illuminated pixel per pulse in order to scan the probability-of-seeing curve, the number of photons going through the inactivated LCD dots should be negligible compared to 200. Since $0.35^5 \approx 1/200$, five passes provide for a photon background two orders of magnitude smaller than the stimulus photons. In **Figure 6b** and **c** we show examples of LCD dot patterns that produce various patterns of pixels across the laser beam. For example, a single pixel is created by a single inactivated dot in the LCD (**Figure 6b**), while the dot arrangement for a 3×3 grid of pixels is shown in **Figure 6c**. For the moment [5] we can illuminate any pixel arrangement in a grid of 5×5 pixels, each about 1 mm width.

In **Figure 6d** we show that indeed the photon statistics of the stimulus light at 532 nm are Poissonian. In particular, this is accomplished by the aforementioned intensity feedback, without which the photon number distribution is wider than the Poissonian. In **Figure 6e** we show that for photon numbers at least equal to 200 the variance of the photon number is equal to the mean photon number per pulse, hence our stimulus light exhibits Poissonian statistics for all photon numbers of interest for the biometrics protocol. It should also be noted that the control over the number of photons, that is, the ability to change the mean number of photons per illuminated pixel per pulse resides in the feedback system used to stabilize the stimulus light. By changing a voltage within the feedback system, we can scan the number of photons, for example, from 20 to 200 photons.

Finally, we discuss the role of the infrared light. The infrared light is used for pointing, that is, for providing information on the geometry of incidence of the stimulus light on the cornea. As can be seen in **Figure 6a**, the laser beam illuminates the eye through a beam splitter, so that the camera sitting behind the beam splitter can image the subject's eye. Moreover, just before the eye we place a glass plate, so that the laser beam is reflected backwards into the camera, since the reflections off the spherical surface of the eye would miss the camera. However, the green stimulus light is too weak (maximum 200 photons per illuminated pixel per pulse) for its reflection to be detected by the camera. Here comes in the infrared light, which is not perceived by the visual system, thus its intensity can be high enough for its reflection to be visible in the camera. This is what is seen in **Figure 6f–h**, where we depict various examples of patterns of pixels incident on the eye. The large bright pixel on the top left part of each image is the reflection of an infrared lamp providing for ambient light for the camera. The other pixels are the infrared reflections of the illuminated pixels of the laser beam. Due to the spatial overlap of the stimulus and the infrared light, these infrared reflections convey the exact position of the stimulating pixels at 532 nm.

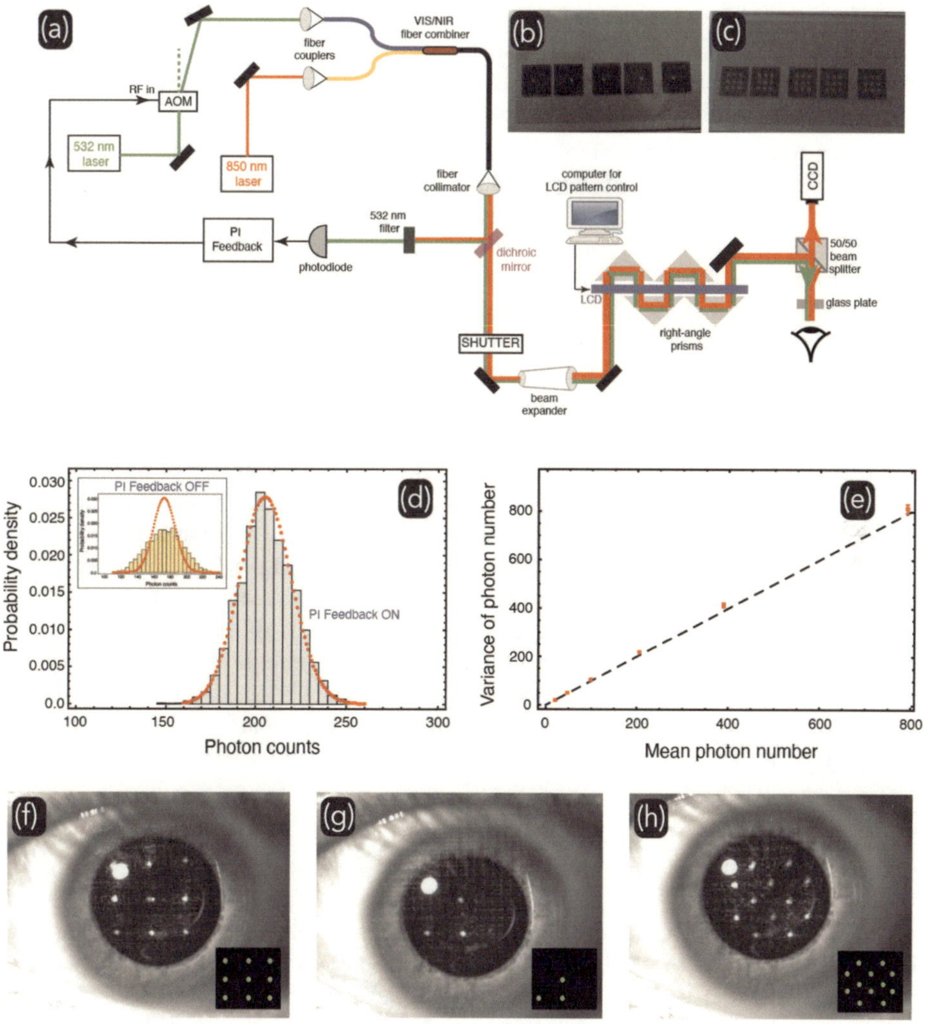

Figure 6.
Optical setup producing a laser beam consisting of an array of pixels, which can be independently illuminated by computer control. The laser beam has two colors combined in a fiber combiner, one at 532 nm used for stimulating the visual system, and the other at 850 nm used as pointing light. (b, c) Pixel patterns are produced by a multi-pass configuration through a liquid crystal display. (d, e) The optoelectronic feedback system stabilizing the intensity of the 532 nm light exiting the fiber combiner leads to Poissonian photon statistics for the time scale and photon number of interest to the interrogation pulsed used in our methodology. (f-h) Examples of various patterns illuminating the eye. What is seen is the reflection of the infrared light off a glass plate before the eye. Reproduced with permission from Springer Nature [5]. Copyright (2020).

7. Quantum advantage with quantum light

One might wonder if there is some advantage to be gained by using quantum light sources for the stimulus light instead of laser light. Indeed, in [4] it was theoretically shown that a single-photon source, for example, a heralded single-photon source [18–21] can lead to a quantum advantage. In particular, it was shown that the total interrogation time is reduced by using single photons. The advantage comes about because the narrower distribution of the incident photon number affects the probabilities p_H and p_L introduced in Section 3, which then reduce the value of the parameter u. This leads to an increase of the probability, P_A, that Alice responds correctly in a single interrogation. Finally, this increase in P_A leads to a smaller number of interrogations required to achieve the same p_{fn} and p_{fp}.

The fact that we can use a single-photon source producing a number of, for example, 200 photons in a light pulse stimulating the visual system rests on the rather large temporal summation window [22], which is the time span within which the visual system cannot temporally resolve the perceived light. Were that not the case, one would need Fock states with up to 200 photons, which so far cannot be produced. In contrast, a heralded-single photon source working at 1 kHz rate would do.

It is interesting to note that the quantum advantage obtained, that is, the required number of required interrogations, is reduced by slightly more than 10% compared to laser light. This figure is at first sight not significant, the main reason being that the statistics of the detected photons differ only slightly [4] between quantum light and laser light, because of the high optical losses suffered by light. It is actually these losses that we take advantage of to define the fingerprint of this method. Since these losses are rather large (typical values of $\alpha \approx 0.1$), the photon statistics of quantum light are "degraded" to the Poissonian statistics. Yet in [4] we provided only the first such proof-of-principle. It is conceivable that different authentication protocols could result in a larger advantage, especially because the visual system is highly nonlinear. This nonlinearity could be used in different ways to amplify the small difference in the photons statistics of detected photons between quantum light and laser light.

8. Conclusions

We have elaborated on a new biometric authentication method, which is based on the human visual system's ability to perform photon counting. The method works with weak light, in order for the effect of visual perception to take place when the light intensity is close to the visual threshold. In such a regime, optical losses suffered by light when propagating from the cornea to the retina are crucial in determining the outcome of perception of weak light flashes. These losses form the biometric "fingerprint" of our biometric authentication methodology. We have described an intuitive authentication algorithm based on illuminating a number of retinal spots being associated with either high optical losses or low optical losses, and used this algorithm to discuss basic features of our methodology, like aging effects, and the fingerprint's inter-subject and intra-subject variability.

We then reviewed recent experimental progress towards developing a laser light stimulus source which provides for light patterns with the desired properties needed for the realization of the authentication protocols. Finally, we presented recent work in exploring a possible quantum advantage that could be obtained by using a quantum light source instead, like a heralded single-photon source.

From a broader perspective, this work further demonstrates the scientific potential of the emerging field of quantum vision, that is, the possibilities for exploring the human and animal visual system using modern photonic and quantum-optical tools [23–28].

Notes/thanks/other declarations

IK and ML acknowledge co-financing of this work by the European Union and Greek national funds through the Operational Program Competitiveness, Entrepreneurship and Innovation, under the call "RESEARCH-CREATE- INNOVATE," with project title "Photonic analysis of the retina's biometric photo-absorption" (project code: T1EDK-04921). OEM acknowledges financial support from the Scientific and Technological Research Council of Turkey (TÜBITAK), grant No. 120F200.

Author details

Iannis Kominis[1,2,3*], Michail Loulakis[4,5] and Özgur E. Müstecaplıoğlu[6]

1 Department of Physics, University of Crete, Heraklion, Greece

2 Institute of Theoretical and Computational Physics, University of Crete, Heraklion, Greece

3 Quantum Biometronics PC, Heraklion, Greece

4 School of Applied Mathematical and Physical Sciences, National Technical University of Athens, Athens, Greece

5 Institute of Applied and Computational Mathematics, Foundation for Research and Technology, Heraklion, Greece

6 Department of Physics, Koç University, Istanbul, Turkey

*Address all correspondence to: ikominis@physics.uoc.gr

References

[1] Loulakis M, Blatsios G, Vrettou CS, Kominis IK. Quantum biometrics with retinal photon counting. Physical Review Applied. 2017;**8**:044012. DOI: 10.1103/PhysRevApplied.8.044012

[2] Gisin N, Ribordy G, Tittel W, Zbinden H. Quantum cryptography. Reviews of Modern Physics. 2002;**74**: 145. DOI: 10.1103/RevModPhys.74.145

[3] Pirandola S et al. Advances in quantum cryptography. Advances in Optics and Photonics. 2019;**12**:1012. DOI: 10.1364/AOP.361502

[4] Kominis IK, Loulakis M. Quantum advantage in biometric authentication with single photons. Journal of Applied Physics. 2022;**131**:084401. DOI: 10.1063/5.0080942

[5] Margaritakis A, Anyfantaki G, Mouloudakis K, Gratsea A, Kominis IK. Spatially-selective and quantum-statistics-limited light stimulus for retina biometrics and pupillometry. Applied Physics B. 2020;**126**:99. DOI: 10.1007/s00340-020-07438-z

[6] Hecht S, Shlaer S, Pirenne MH. Energy, quanta and vision. Journal of General Physiology. 1942;**25**:819. DOI: 10.1085/jgp.25.6.819

[7] Bialek W. Biophysics: Searching for Principles. Princeton: Princeton University Press; 2012. p. 656

[8] Tinsley JN, Molodtsov MI, Prevedel R, Wartmann D, Espigul'e-Pons J, Lauwers M, et al. Direct detection of a single photon by humans. Nature Communications. 2016;**7**:12172. DOI: 10.1038/ncomms12172

[9] Baylor BA, Lamb TD, Yau KW. Responses of retinal rods to single photons. Journal of Physiology. 1979;**288**: 613. DOI: 10.1113/jphysiol.1979.sp012716

[10] Rieke F, Baylor DA. Origin of reproducibility in the responses of retinal rods to single photons. Biophysical Journal. 1998;**75**:1836. DOI: 10.1016/S0006-3495(98)77625-8

[11] Rieke F, Baylor DA. Single-photon detection by rod cells of the retina. Reviews of Modern Physics. 1998;**70**:1027. DOI: 10.1103/RevModPhys.70.1027

[12] Sim N, Bessarab D, Jones CM, Krivitsky LA. Method of targeted delivery of laser beam to isolated retinal rods by fiber optics. Biomedical Optics Express. 2011;**2**:2926. DOI: 10.1364/BOE.2.002926

[13] Sim N, Cheng MF, Bessarab D, Jones CM, Krivitsky LA. Measurement of photon statistics with live photoreceptor cells. Physical Review Letters. 2012;**109**: 113601. DOI: 10.1103/PhysRevLett.109.113601

[14] Phan NM, Cheng MF, Bessarab DA, Krivitsky LA. Interaction of fixed number of photons with retinal rod cells. Physical Review Letters. 2014;**112**: 213601. DOI: 10.1103/PhysRevLett.112.213601

[15] Nelson PC. Old and new results about single-photon sensitivity in human vision. Physical Biology. 2016;**13**:025001. DOI: 10.1088/1478-3975/13/2/025001

[16] Racette L, Fisher M, Bebie H, Holló G, Johnson CA, Matsumoto C. Visual Field Digest. Switzerland: Haag-Streit AG; 2019

[17] Heijl A, Lindgren G, Olsson J. Normal variability of static perimetric

threshold values across the central visual field. Archives of Ophthalmology. 1987; **105**:1544. DOI: 10.1001/archopht.1987.01060110090039

[18] Oxborrow M, Sinclair AG. Single-photon sources. Contemporary Physics. 2005;**46**:173. DOI: 10.1080/00107510512331337936

[19] Buller GS, Collins RJ. Single-photon generation and detection. Measurement Science and Technology. 2010;**21**: 012002. DOI: 10.1088/0957-0233/21/1/012002

[20] Eisaman MD, Fan J, Migdall A, Polyakov SV. Invited review article: Single-photon sources and detectors. Review of Scientific Instruments. 2011; **82**:071101. DOI: 10.1063/1.3610677

[21] Meyer-Scott E, Silberhorn C, Migdall A. Single-photon sources: Approaching the ideal through multiplexing. Review of Scientific Instruments. 2020;**91**: 041101. DOI: 10.1063/5.0003320

[22] Holmes R, Victora M, Wang RF, Kwiat PG. Measuring temporal summation in visual detection with a single-photon source. Vision Research. 2017;**140**:33. DOI: 10.1016/j.visres.2017.06.011

[23] Brunner N, Branciard C, Gisin N. Possible entanglement detection with the naked eye. Physical Review A. 2008;**78**: 052110. DOI: 10.1103/PhysRevA.78.052110

[24] Lucas F, Hornberger K. Incoherent control of the retinal isomerization in rhodopsin. Physical Review Letters. 2014;**113**:058301. DOI: 10.1103/PhysRevLett.113.058301

[25] Pizzi R, Wang R, Rossetti D. Human visual system as a double-slit single photon interference sensor: A comparison between modellistic and biophysical tests. PLoS One. 2016;**11**: e0147464. DOI: 10.1371/journal.pone.0147464

[26] Dodel A, Mayinda A, Oudot E, Martin A, Sekatski P, Bancal JD, et al. Proposal for witnessing non-classical light with the human eye. Quantum. 2017;**1**:7. DOI: 10.22331/q-2017-04-25-7

[27] Sarenac D, Kapahi C, Silva AE, Cory DG, Taminiau I, Thompson B, et al. Direct discrimination of structured light by humans. Proceedings of the National Academy of Sciences USA. 2020;**117**: 14682. DOI: 10.1073/pnas.1920226117

[28] Pedram A, Müstecaplıoğlu ÖE, Kominis IK. Using quantum states of light to probe the retinal network. arXiv: 2111.03285.

Chapter 8

Feature Extraction Using Observer Gaze Distributions for Gender Recognition

Masashi Nishiyama

Abstract

We determine and use the gaze distribution of observers viewing images of subjects for gender recognition. In general, people look at informative regions when determining the gender of subjects in images. Based on this observation, we hypothesize that the regions corresponding to the concentration of the observer gaze distributions contain discriminative features for gender recognition. We generate the gaze distribution from observers while they perform the task of manually recognizing gender from subject images. Next, our gaze-guided feature extraction assigns high weights to the regions corresponding to clusters in the gaze distribution, thereby selecting discriminative features. Experimental results show that the observers mainly focused on the head region, not the entire body. Furthermore, we demonstrate that the gaze-guided feature extraction significantly improves the accuracy of gender recognition.

Keywords: gaze distribution, region of interest, feature extraction, pedestrian image, gender recognition

1. Introduction

Gender recognition, which is of interest in the field of soft-biometrics, is part of the collection of statistical data about people in public spaces. Furthermore, gender recognition has many potential applications, such as video surveillance and consumer behavior analysis. Often, gender recognition experiments are conducted on pedestrians captured on video. Researchers have proposed several methods for automatically recognizing gender in pedestrian images; many of these techniques use convolutional neural networks (CNNs) [1]. The existing methods can extract discriminative features for gender recognition and obtain highly accurate results when many training samples containing diverse pedestrian images are acquired in advance. However, the collection of a sufficient number of training samples is very time-consuming. Unfortunately, deep learning methods typically require these large training sets to maintain suitable recognition performance.

People quickly and correctly recognize gender; thus, we believe that people effectively extract visual features from subjects in images. For instance, people correctly recognize gender from facial images [2, 3]. It may be possible to reproduce human visual abilities in a computer algorithm with a small number of training samples and achieve a recognition performance equivalent to that of humans. Existing methods [4, 5] have been proposed to mimic human visual abilities for object recognition tasks. These methods used a saliency map generated from low-level features [6–8]. However, these saliency maps does not sufficiently represent human visual abilities because they are not directly measured from human observers. We thus consider that the existing methods disregard the deep mechanisms of human vision.

An increasing number of pattern recognition studies, specifically those attempting to mimic human visual ability, have measured the gaze distribution of observers [9–12]. This gaze distribution has great potential in the collection of informative features for various recognition tasks. Several techniques [13, 14] have demonstrated that the gaze distribution facilitates the extraction of informative features. Sattar et al. [13] applied the gaze distribution to analyze fashion in images. Murrugarra-Llerena and Kovashka [14] applied the gaze distribution for attribute prediction in facial images. However, the existing methods using observer gaze distribution do not study gender recognition from pedestrian images. We consider that the region of interest measured from observers' gaze is also effective for gender recognition.

Here, we conduct a gaze measurement experiment for observers performing a gender recognition task on images of subjects. We investigate if the gaze distribution measured from the observers facilitates gender recognition. **Figure 1** shows the overview of our gaze-guided feature extraction. We generate a task-oriented gaze distribution from the gaze locations recorded while observers manually determined the genders of subjects in images. High values in a task-oriented gaze distribution correspond to regions that observers frequently view. We assume that these regions contain discriminative features for gender recognition because they appear to be useful when

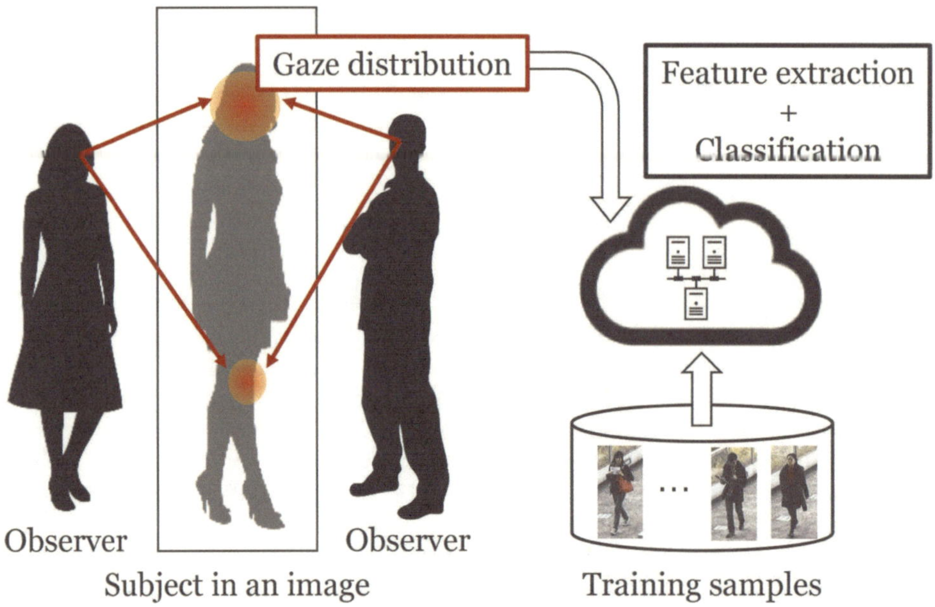

Figure 1.
Overview of our gaze-guided feature extraction. We consider that the regions gathering the gaze distribution contain discriminative features for gender recognition because they appear to be useful when the observers are tackling the gender recognition task.

the observers are determining the subject gender. When extracting features to train the gender classifier, larger weights are assigned to the regions of the pedestrian images corresponding to the attention regions of the task-oriented gaze distribution. The experimental results indicate that our gaze-guided feature extraction improves the gender recognition accuracy when using a CNN technique with a small number of training samples.

2. Generating a task-oriented gaze distribution for gender recognition

2.1 Observer gaze distribution in gender recognition

We discuss which body regions of subjects in images are frequently viewed for gender recognition by observers. With respect to the analytical study of facial images, Hsiao et al. [15] reported that people looked at the nose region when they recognized others. We consider that the human face is a key factor in gender recognition. Furthermore, we consider that the entire body, including the chest, waist, and legs, is also helpful. Thus, we aim to reveal the body regions that tend to collect the observer gaze distribution during a gender recognition task. Note that we assume that the pedestrian images have been pre-aligned using pedestrian detection techniques. The details of our method are described below.

2.2 Generating a task-oriented gaze distribution

To generate a task-oriented gaze distribution, we use a gaze tracker to acquire gaze locations while the observer views a pedestrian image on a screen. We briefly describe our method in **Figure 2**. We work with P participating observers and N pedestrian images. Given a gaze location $\left(x_f, y_f\right)$ in a certain frame f, the gaze distribution $g_{p,n,f}(x,y)$ is computed as

$$g_{p,n,f}(x,y) = \begin{cases} 1 & \left(x = x_f, y = y_f\right), \\ 0 & (\text{otherwise}), \end{cases} \tag{1}$$

where p is an observer, and n is a pedestrian image. Note that the observer not only looks at point $\left(x_f, y_f\right)$ on each pedestrian image, but also the region surrounding this point. Thus, we apply a Gaussian kernel to the measured gaze distribution $g_{p,n,f}(x,y)$. **Figure 3** illustrates the parameters used to determine the size k of the Gaussian kernel. We compute the following equation:

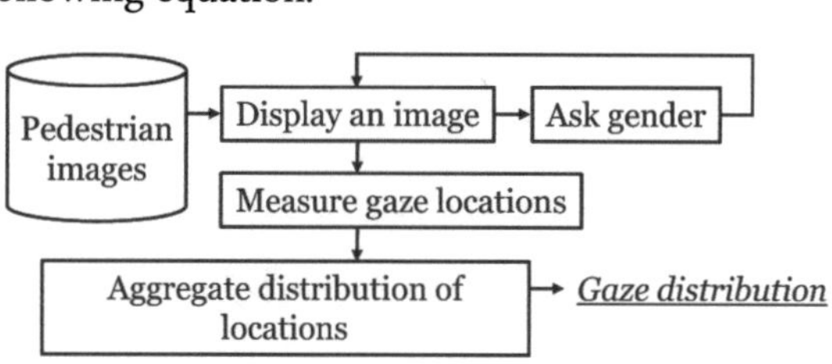

Figure 2.
Overview of our method for generating a gaze distribution $\tilde{g}(x,y)$.

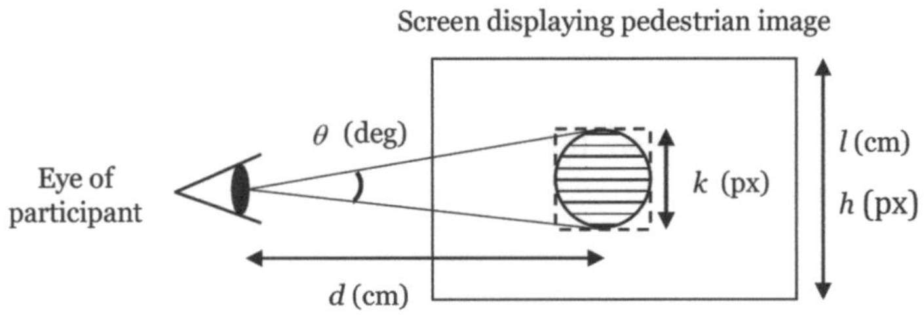

Figure 3.
Parameters used to determine the kernel size for generating the gaze distribution.

$$k = \frac{2dh}{l} \tan \frac{\theta}{2}, \tag{2}$$

where θ represents the angle of the region surrounding $\left(x_f, y_f\right)$, l represents the screen's vertical length, h represents the screen's vertical resolution, and d represents the distance from the participant to the screen. We aggregate each $g_{p,n,f}(x,y)$ to $g_{p,n}(x,y)$ to represent the gaze distribution in a particular pedestrian image as

$$g_{p,n}(x,y) = \sum_{f=1}^{F_{p,n}} k(u,v) * g_{p,n,f}(x,y), \tag{3}$$

where $F_{p,n}$ is the time taken by an observer to recognize the gender of the subject in the image. Function $k(u,v)$ represents a Gaussian kernel of size $k \times k$ and operator $*$ represents the convolution. Our method performs L1-norm normalization as $\|g_{p,n}(x,y)\| = 1$. We aggregate $g_{p,n}(x,y)$ into a single gaze distribution across all observers and all pedestrian images. The aggregated gaze distribution $g(x,y)$ is represented as

$$g(x,y) = \sum_{p=1}^{P} \sum_{n=1}^{N} g_{p,n}(x,y). \tag{4}$$

Note that we apply a scaling technique to the aggregated gaze distributions as follows: $\tilde{g}(x,y) = g(x,y)/\max(g(x,y))$. $\tilde{g}(x,y)$ is the final task-oriented gaze distribution.

3. Experiments to generate a task-oriented gaze distribution

3.1 Setup

We evaluated the task-oriented gaze distributions for gender recognition. We acquired the gaze locations for $P = 14$ participating observers (seven men and seven women, with an average age of 22.6 ± 1.3 years old, Japanese students). We used a display screen (size 53.1×29.9 cm, 1920×1080 pixels). The vertical distance between the screen and the participant was set to 65 cm, as illustrated in **Figure 4**.

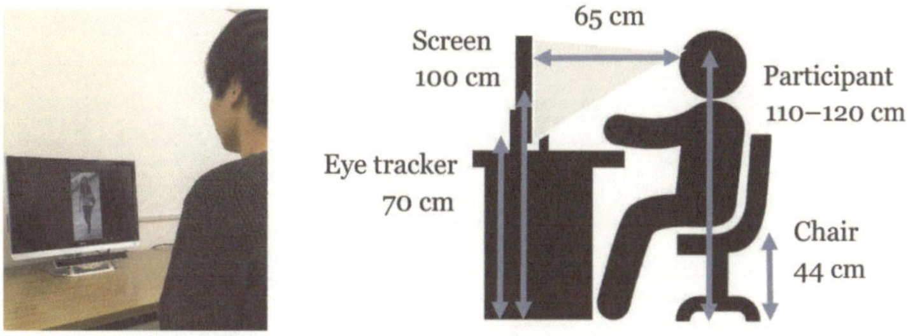

Figure 4.
Setup used to acquire the gaze distribution in a gender recognition task.

The height from the floor to the eyes of the participant was between 110 cm and 120 cm. The participants sat on a chair in a room with no direct sunlight (illuminance 825 lx). We use a standing eye tracker (GP3 Eye Tracker, sampling rate 60 Hz). We asked the participants to perform a gender recognition task to determine if the pedestrian in an image is a man or a woman. We determined which regions of the entire body were viewed by the participants to complete this task.

We used 4563 pedestrian images from the CUHK dataset included in the PETA dataset [16] with gender labels (woman or man). From this dataset, we use the $N = 8$ pedestrian images in **Figure 5** to use in the observer experiment to generate the gaze distribution map. We selected the four pedestrian images at the top of **Figure 5** keeping the ratio of directions (front, back, left, and right) equal. We selected the remaining pedestrian images in **Figure 5** in the same manner. When displaying the stimulus images on the screen, the pedestrian images were enlarged from 80×160 pixels to 480×960 pixels. We simply changed the stimulus images' positions by adding random offsets to avoid a center bias [17, 18].

We acquired the gaze distribution when participants performed the gender recognition task according to the following procedure:

P1. A gray image is shown on the screen for one second.

P2. A pedestrian stimulus image is shown on the screen for two seconds.

P3. A black image is shown on the screen for two seconds, and the participant replied whether the pedestrian was a woman or a man.

P4. We repeated P1 to P3 until all eight pedestrian images had been displayed in random order.

In our preliminary experiment, we observed that participants first assessed the position of the pedestrian image on the screen and then, after establishing the position of the image, attempted to complete the gender recognition task. To determine $F_{p,n}$, we set the start time at the point when the gaze first stopped on the pedestrian image for more than 440 ms, and the end time corresponds to the pedestrian image no longer appearing on screen. In this scenario, the average $F_{p,n}$ between the start and end times was 1.56 ± 0.38 s. The participating observers achieved a gender recognition was accuracy of 100.0%.

We set $\theta = 3°$ in Eq. (2) by considering the range of the fovea, which is approximately two degrees (as described in [19]), and the error of the eye tracker, which is about one degree (as described in the tracker's specification sheet). We used a kernel size of $k = 125$ for the enlarged pedestrian images (480×960 pixels). The size of the

Figure 5.
Pedestrian images for generating task-oriented gaze distributions during the gender recognition task.

gaze distribution images was downsized by 80 × 160, adjusting according to the original size of the pedestrian images. This standardized the size of the test samples and training samples to input into the gender classifier.

3.2 Results

Figure 6 shows examples of the measured gaze distributions $g_{p,n}(x, y)$ for the gender recognition task for a pedestrian image. We show the gaze distribution map from two participants for the pedestrian image shown in **Figure 6(a)**. The dark regions in the gaze distribution maps represent the gaze locations recorded from the participants by the eye tracker. The minimum (black) and maximum (white) intensities in **Figure 6** represent the maximum and minimum values of the measured $g_{p,n}(x, y)$, respectively. We observed that participants frequently concentrated their gaze on the head region to complete the gender recognition task.

Figure 7 shows the overall task-oriented gaze distribution $\tilde{g}(x, y)$ for gender recognition synthesized from all of the participating observers. To study the properties of the task-oriented gaze distribution, we verify how the gaze distributions align with the pedestrian images of **Figure 5**. We see that the region corresponding to the head

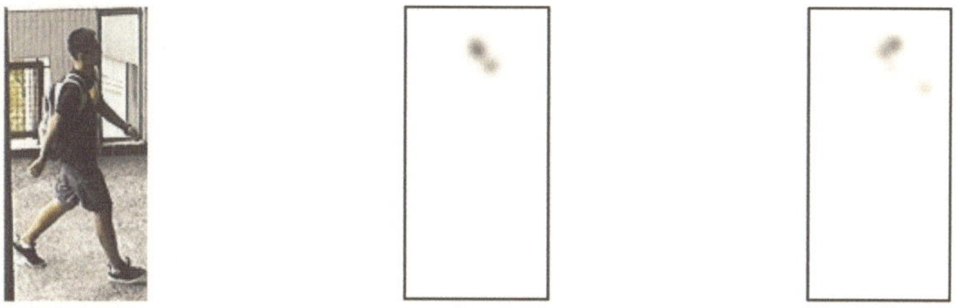

(a) Stimulus image of pedestrian (b) Measured gaze distribution of participant 1 (c) Measured gaze distribution of participant 2

Figure 6.
Examples of measured gaze distributions $g_{p,n}(x, y)$ from two participants. (a) Stimulus image of pedestrian. (b) and (c) Gaze distributions measured from each participant viewing the pedestrian image in (a).

Figure 7.
Task-oriented gaze distribution $\tilde{g}(x, y)$ for the gender recognition task.

gathered a large number of gaze locations, while regions around the lower body and background gathered few gaze locations.

4. Feature extraction algorithm using the task-oriented gaze distribution for gender recognition

4.1 Overview of our gaze-guided feature extraction

Here, we describe our method to extract features using the task-oriented gaze distribution for gender recognition. The regions corresponding to high values in the distribution $\tilde{g}(x, y)$ appear to contain informative features because participants focus on these regions to manually recognize gender in the pedestrian images. Thus, we assume that these regions contain discriminative features for the gender classifiers. Based on this assumption, we extract these features by assigning higher weights to the regions corresponding to high values in the task-oriented gaze distribution. **Figure 8** provides an overview of our method. Our methods assign weights using $\tilde{g}(x, y)$ for both the test samples and training samples. Therefore, we do not need to

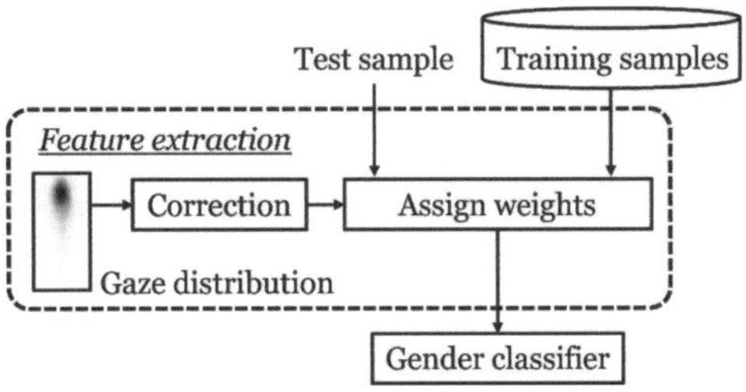

Figure 8.
Overview of our gaze-guided feature extraction using the gaze distribution $\tilde{g}(x,y)$.

acquire gaze distributions on the test samples. Our method extracts the weighted features and applies deep learning and machine learning techniques to obtain the final classification.

4.2 Procedure

Given a gaze distribution $\tilde{g}(x,y)$, our method computes the weight $\tilde{w}(x,y)$ for each pixel as

$$\tilde{w}(x,y) = C(\tilde{g}(x,y)). \tag{5}$$

We use a correction function C() that weakens or emphasizes values according to the density of the gaze distribution.

We calculate a weighted intensity $i_w(x,y)$ from an original intensity $i(x,y)$ as follows:

$$i_w(x,y) = \tilde{w}(x,y)i(x,y). \tag{6}$$

We generate a feature vector for gender recognition using raster scanning $i_w(x,y)$. The RGB images are converted to CIE L*a*b* color space. Note that our method weights the L* values and does not change the a*b* values. We consider only the lightness changes without any color changes because a numerical change in the L* channel corresponds to the lightness change in human perception.

5. Evaluation of the gender recognition performance using the gaze distribution

5.1 Comparison of weight correction functions for feature extraction

We evaluated the accuracy of gender recognition using various correction functions. We used the gaze distribution $\tilde{g}(x,y)$, as shown in **Figure 7**. We randomly picked up pedestrian images from the CUHK dataset, which is included in the PETA dataset [16]. We equalized the ratio of women and men samples in the test sets and training sets to avoid problems associated with imbalanced data. The same individual did not appear in both training and test samples. We used 2720 pedestrian images as training samples and test samples. We applied 10-fold cross-validation for gender recognition. Both the training and test samples contained not only frontal poses, but also side and back poses. We evaluate the gender recognition performance as the accuracy of the woman or man classification labels. We generated feature vectors by raster scanning RGB values with down sampling ($40 \times 80 \times 3$ dimensions) from weighted pedestrian images. We used a linear support vector machine classifier (the penalty parameter was $C = 1$) to confirm the baseline performance of gender recognition. For the other classifiers, we show experimental results in Section 5.2. We compared the accuracy of the following correction functions:

F1. $\mathrm{C}(z) = z$,

F2. $\mathrm{C}(z) = \min\{1, z^a + b\}$,

F3. $\mathrm{C}(z) = 1 - \min\{1, z^a + b\}$, and

F4. $\mathrm{C}(z) = 1$.

Figure 9 shows a visualization of the correction functions $\mathrm{C}(z)$. We determined the parameters of the gender classifier using a grid search over the validation sets. These validation sets consisted of the remaining pedestrian images not used in the test sets and training sets from the CUHK dataset. Parameters $\{a, b\}$ were set to $\{0.75, 0.21\}$.

Figure 10(a) shows pedestrian images after applying $\mathrm{C}(z)$. Function F1 outputs an intensity weighted by the gaze distribution for each pixel. Function F2 emphasizes an intensity around a face using gaze distribution. In contrast, function F3 weakens the intensity. Function F4 directly outputs the intensity of the original pedestrian image.

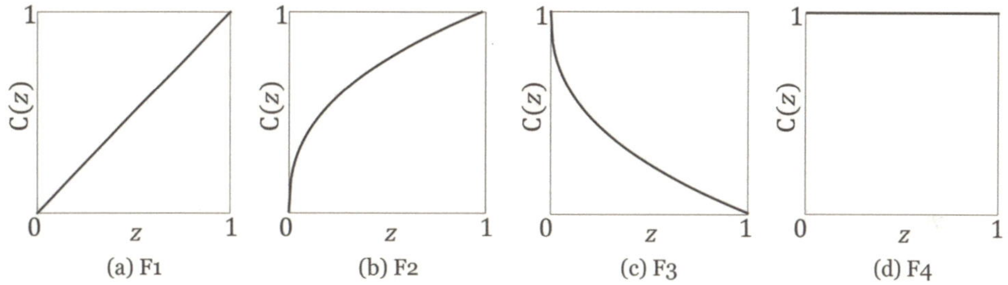

Figure 9.
Visualization of correction functions $\mathrm{C}(z)$.

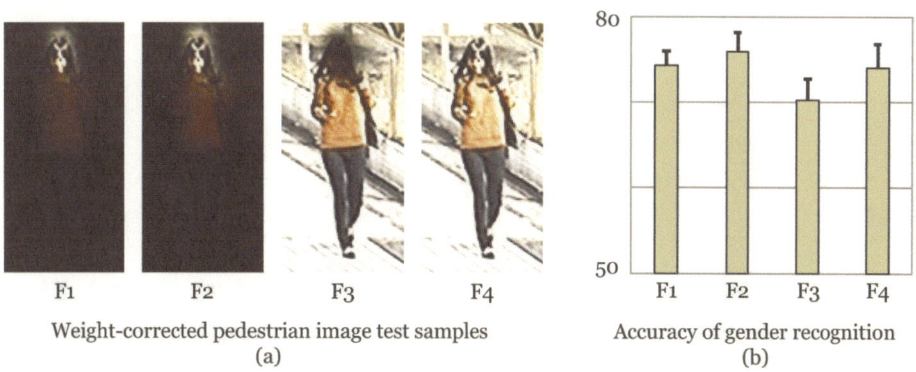

Figure 10.
Gender recognition accuracy. (a) Examples of test pedestrian images after applying correction functions. F1 and F2 show the results of our gaze-based feature extraction. (b) Comparison of gender recognition accuracy using each gaze-guided weight correction function with a linear support vector machine classifier.

Figure 10(b) shows the gender recognition accuracy of each gaze-guided weight correction function for gender recognition. We confirmed that the accuracy of F1 and F2 was superior to that of F4. Thus, the use of the gaze distribution $\tilde{g}(x,y)$ appears to increase the performance of gender recognition. F2 yields superior performance compared with F1, indicating that this correction function improves gender recognition accuracy. The inverse weights of F3 decreased the accuracy compared with the other correction functions. Thus, we demonstrate that the regions corresponding to observer gaze distribution $\tilde{g}(x,y)$ measured from participants completing a gender recognition task contain discriminative features for the gender classifier.

5.2 Combining our gaze-guided feature extraction with existing classifiers

We investigated the gender recognition performance by combining our gaze-based feature extraction technique with representative classifiers. We used *Mini*-CNN architecture [20], which is a small network with few convolutional layers. We also used a large margin nearest neighbor (LMNN) classifier [21], which is a metric learning technique. The test samples and training samples described in Section 5.1 were used in the evaluation. We applied 10-fold cross-validation. **Table 1** shows the accuracy for gender recognition with and without our gaze-guided feature extraction. Our gaze-based feature extraction method leads to improved gender recognition for both classifiers.

5.3 Evaluation of assigning weights using saliency maps

We evaluated the gender recognition accuracy of a method that uses saliency maps. We used the existing methods of Zhang et al. [7], and Zhu et al. [8] to generate saliency maps. **Figure 11** shows the saliency maps used in the evaluation of gender recognition. We scaled the intensity in the saliency map to fit the intensity range to [0,1]. We performed feature extraction using the saliency map instead of the task-

Condition	Accuracy using CNN	Accuracy using LMNN
With our gaze-guided feature extraction	79.6 ± 2.2	78.5 ± 1.1
Without our gaze-guided feature extraction	75.3 ± 3.1	76.0 ± 2.7

Table 1.
Accuracy (%) of gender recognition by combining our gaze-guided feature extraction with existing classifiers.

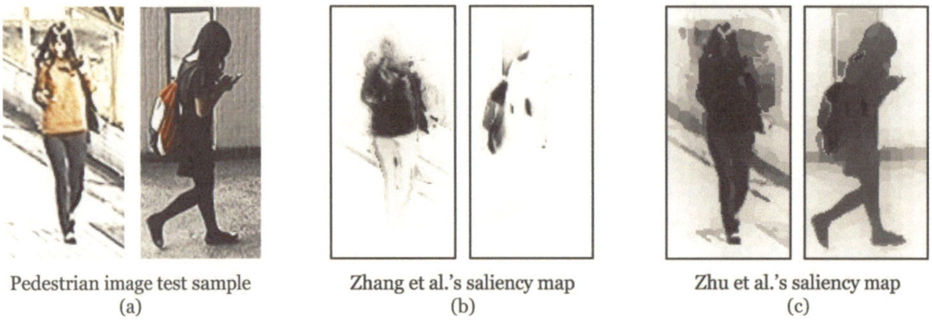

Figure 11.
Examples of saliency maps used in gender recognition. (a) Test pedestrian images. (b), (c) generated saliency maps.

Our gaze distribution	Zhang et al.'s saliency map	Zhu et al.'s saliency map
79.6 ± 2.2%	66.9 ± 2.5%	66.8 ± 2.8%

Table 2.
Gender recognition accuracy (%) using our task-oriented gaze distribution compared with using the existing saliency maps.

oriented gaze distribution $\tilde{g}(x,y)$. Our method assigned the test samples and training samples large weights in regions corresponding to high saliency values before using a CNN classifier. We evaluated the accuracy using the same conditions of Section 5.2. **Table 2** shows the gender recognition accuracy obtained when using our task-oriented gaze distribution compared with the accuracy obtained using the existing saliency map approaches. The results indicate that our gaze-guided feature extraction method outperforms the use of saliency maps for gender recognition.

5.4 Visualization of the regions of focus when using CNNs

We conducted an experiment to visualize the regions of focus in a pedestrian image during gender recognition. To this end, we used gradient-weighted class activation mapping (Grad-CAM) [22]. **Figure 12** shows the visualization results of the regions of focus of the CNN method. In (a), we show the pedestrian test images for gender recognition. In (b), we show the visualization results without our gaze-guided feature extraction. We only used the conventional CNN of the VGG16 model with fine-tuning. In the woman test samples, the model emphasized the leg and waist regions. In the man test samples, the model emphasized the shoulder and head regions. This indicates that the conventional CNN emphasizes various body part regions for gender recognition but in a different manner than used by the participating observers in the experiments of Section 3.2. In (c), we show the visualization results using our gaze distribution maps for gender recognition. We used our gaze-guided feature extraction with *Mini*-CNN, as described in Section 5.2. We confirmed that our method mainly emphasizes the head region, mimicking the human observers' gaze behavior. In particular, we consider that our method recognizes gender by focusing on the hairstyle of the subject in an image because it emphasized the regions containing the boundary between the head and the background.

6. Conclusions

We hypothesized that the gaze distribution measured from observers performing a gender recognition task facilitates the extraction of discriminative features. We demonstrated that the gaze distribution measured during a manual gender recognition task tended to concentrate on specific regions of the pedestrian's body. We represented the informative region as a task-oriented gaze distribution for a gender classifier. Owing to the efficacy of the task-oriented gaze distribution for feature extraction, our gender recognition method demonstrated increased accuracy compared with representative existing classifiers and saliency maps.

As part of our future work, we will expand our analytical study to explore the differences in gaze distributions with respect to observer nationality and ethnicity. Furthermore, we intend to generate gaze distributions for various tasks beyond gender recognition, such as evaluating impressions of subjects' clothing in images.

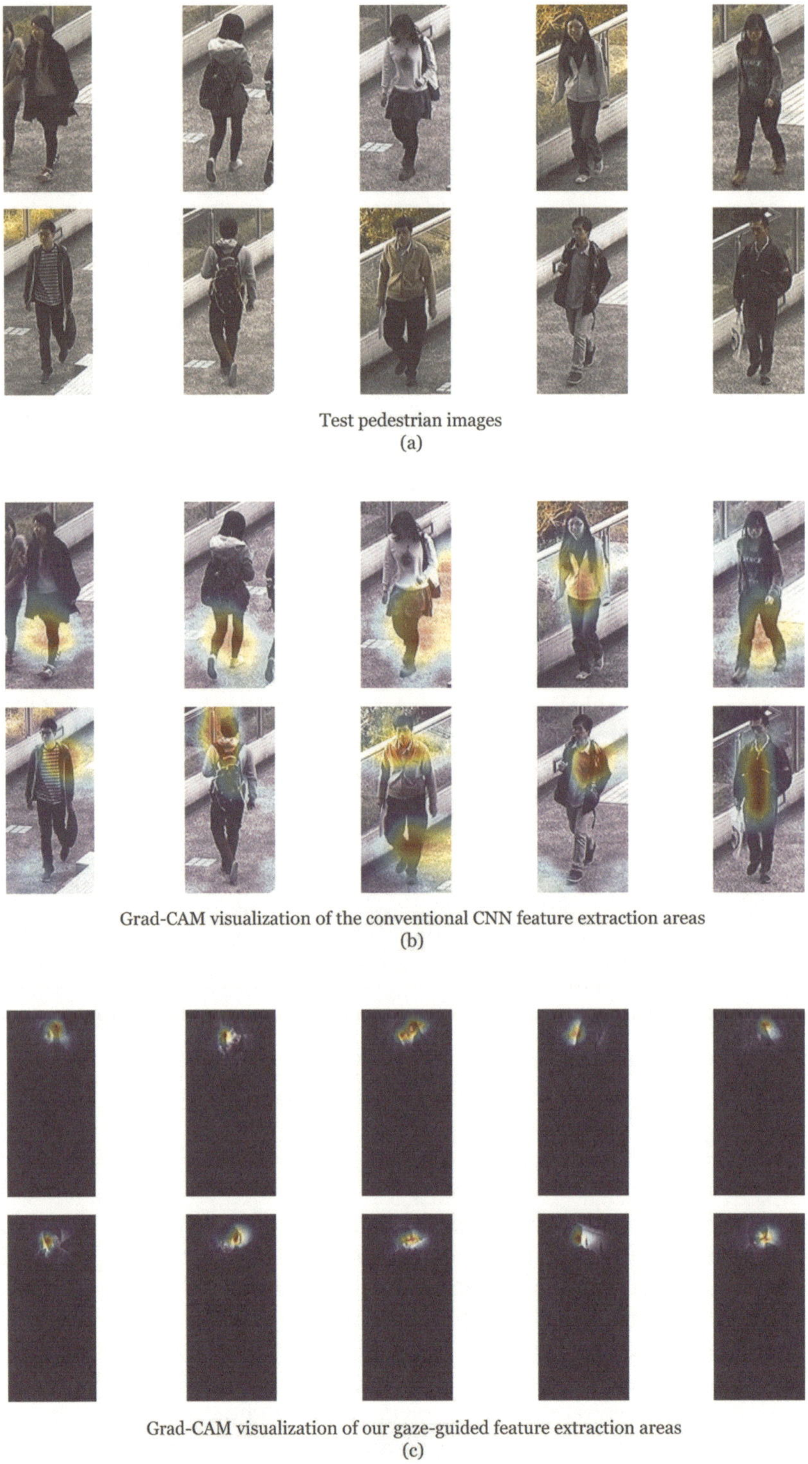

Test pedestrian images
(a)

Grad-CAM visualization of the conventional CNN feature extraction areas
(b)

Grad-CAM visualization of our gaze-guided feature extraction areas
(c)

Figure 12.
Regions of focus of the gender classifier when performing gender recognition. We used CNNs and Grad-CAM. (a) Test pedestrian images. (b) Results without the use of the gaze distribution $\tilde{g}(x, y)$. (c) Results with our gaze-guided feature extraction.

Acknowledgements

This work was partially supported by JSPS KAKENHI Grant No. JP20K11864.

Author details

Masashi Nishiyama
Graduate School of Engineering, Tottori University, Tottori, Japan

*Address all correspondence to: nishiyama@tottori-u.ac.jp

References

[1] Fayyaz M, Yasmin M, Sharif M, Raza M. J-ldfr: Joint low-level and deep neural network feature representations for pedestrian gender classification. Neural Computing and Applications. 2021;**33**:361-391

[2] Bruce V, Burton AM, Hanna E, Healey P, Mason O, Coombes A, et al. Sex discrimination: How do we tell the difference between male and female faces? Perception. 1993;**22**(2):131-152

[3] Burton AM, Bruce V, Dench N. What's the difference between men and women? Evidence from facial measurement. Perception. 1993;**22**(2):153-176

[4] Walther D, Itti L, Riesenhuber M, Poggio T, Koch C. Attentional selection for object recognition—A gentle way. In: Proceedings of the Second International Workshop on Biologically Motivated Computer Vision. Berlin Heidelberg: Springer; 2002. pp. 472-479

[5] Zhu JY, Wu J, Xu Y, Chang E, Tu Z. Unsupervised object class discovery via saliency-guided multiple class learning. IEEE Transactions on Pattern Analysis and Machine Intelligence. 2015;**37**(4): 862-875

[6] Itti L, Koch C, Niebur E. A model of saliency-based visual attention for rapid scene analysis. IEEE Transactions on Pattern Analysis and Machine Intelligence. 1998;**20**(11):1254-1259

[7] Zhang J, Sclaroff S, Lin X, Shen X, Price B, Mech R. Minimum barrier salient object detection at 80 fps. In: Proceedings of the IEEE International Conference on Computer Vision. IEEE Computer Society; 2015. pp. 1404-1412

[8] Zhu W, Liang S, Wei Y, J. sun. Saliency optimization from robust background detection. In: Proceedings of the IEEE Conference on Computer Vision and Pattern Recognition. IEEE Computer Society; 2014. pp. 2814-2821

[9] Xu M, Ren Y, Wang Z. Learning to predict saliency on face images. In: Proceedings of IEEE International Conference on Computer Vision. IEEE Computer Society; 2015. pp. 3907-3915

[10] Fathi A, Li Y, Rehg JM. Learning to recognize daily actions using gaze. In: Proceedings of the 12th European Conference on Computer Vision. Berlin Heidelberg: Springer; 2012. pp. 314-327

[11] Xu J, Mukherjee L, Li Y, Warner J, Rehg JM, Singh V. Gaze-enabled egocentric video summarization via constrained submodular maximization. In: Proceedings of the IEEE Conference on Computer Vision and Pattern Recognition. IEEE Computer Society; 2015. pp. 2235-2244

[12] Karessli N, Akata Z, Schiele B, Bulling A. Gaze embeddings for zero-shot image classification. In: Proceedings of the IEEE Conference on Computer Vision and Pattern Recognition. IEEE Computer Society; 2017. pp. 4525-4534

[13] Sattar H, Bulling A, Fritz M. Predicting the category and attributes of visual search targets using deep gaze pooling. In: Proceedings of IEEE International Conference on Computer Vision Workshops. IEEE Computer Society; 2017. pp. 2740-2748

[14] Murrugarra-Llerena N, Kovashka A. Learning attributes from human gaze. In: Proceedings of IEEE Winter Conference on Applications of Computer Vision. IEEE Computer Society; 2017. pp. 510-519

[15] Hsiao JH, Cottrell G. Two fixations suffice in face recognition. Psychological Science. 2008;**19**(10):998-1006

[16] Deng Y, Luo P, Loy CC, Tang X. Pedestrian attribute recognition at far distance. In: Proceedings of the 22nd ACM International Conference on Multimedia. Association for Computing Machinery; 2014. pp. 789-792

[17] Bindemann M. Scene and screen center bias early eye movements in scene viewing. Vision Research. 2010;**50**(23): 2577-2587

[18] Buswell GT. How People Look at Pictures: A Study of the Psychology of Perception of Art. Chicago, IL: University of Chicago Press; 1935

[19] Fairchild MD. Color Appearance Models. 3rd ed. New York City: Wiley; 2013

[20] Antipov G, Berrani SA, Ruchaud N, Dugelay JL. Learned vs. hand-crafted features for pedestrian gender recognition. In: Proceedings of the 23rd ACM International Conference on Multimedia. Association for Computing Machinery; 2015. pp. 1263-1266

[21] Weinberger KQ, Saul LK. Distance metric learning for large margin nearest neighbor classification. Journal of Machine Learning Research. 2009;**10**: 207-244

[22] Selvaraju RR, Cogswell M, Das A, Vedantam R, Parikh D, Batra D. Grad-cam: Visual explanations from deep networks via gradient-based localization. In: Proceedings of IEEE International Conference on Computer Vision. IEEE Computer Society; 2017. pp. 618-626

Chapter 9

Image Acquisition for Biometric: Face Recognition

Siddharth B. Dabhade, Nagsen S. Bansod, Yogesh S. Rode, Narayan P. Bhosale, Prapti D. Deshmukh and Karbhari V. Kale

Abstract

Biometrics is mostly used for authentication purposes in security. Due to the covid-19 pandemic situation, nowadays distance-based authentication systems are more focused. Face recognition is one of the best approaches which can use for authentication at distance. Face recognition is a challenging task in various environments. For that taking input from the camera is very important for real-time applications. In this chapter, we are more focusing on how to acquire the face image using MATLAB. The complete chapter is divided into five sections introduction, definition of biometrics, image acquisition devices, image acquisition process in MATLAB.

Keywords: face recognition, biometric, image acquisition, image processing, imtool

1. Introduction

Biometrics is the science of establishing the identity of an individual based on the physical, chemical or behavioral attributes of the person [1, 2]. Those attributes or properties of an individual are unique on the earth called as biometrics identifiers. Physical properties of the person do not vary as per time such as the face, fingerprint, retina, iris, etc. Behavioral biometrics such as voice, signature, and keystroke dynamics identification and measurement of performance of the person while the certain actions of the human through its body parts such as voice-scan and signature-scan. The element of time is essential to behavioral biometrics because it may change with time [3].

In the internet world, there are so many business companies doing their business through client–server basis in which they are authenticating the client's request through the username and password. It may be chances of making an illegal entry in demand and request or access the private and confidential data.

When an end-user uses some additional materials or information for the authentication such as smart card, username, and passwords, some tokens or ids, passport, driving license, etc. then there is a chance of lost, stolen the things you are belonging or passwords, ids may be guessed or forget [4]. Therefore, we required a type of system in which there is no need to use such type of external resources for authentication. Fortunately, a biometric authentication system provides an alternative and robust identification system for these problems. In this system, the user should be present personally at the time of identification or verification. As per security is concerned, it uses three approaches for authentication of the person. The first approach is small text information you know such as password or pin or security questions, etc. The second approach is you are belonging with something such as key, RFID, ATM Card or Smart Card, etc. and the third one is some information is always with you it cannot forget, stolen. Your presence is mandatory for this type of authentication i.e. biometric. Apart from these approaches biometrics is a more suitable system because it is always with the person, therefore, biometrics never borrowed, stolen or forgotten [5].

Biometric is a process of identification of unique patterns from the physical, behavioral or chemical properties of the person for authentication. Face, fingerprint, iris, palm, retina, hand geometry, etc. are physical biometric traits whereas voice, gait, dynamic keystrokes are behavioral and DNA, saliva, body odor, etc. are chemical biometrics traits [6, 7].

The process of how biometric works (shown in **Figure 1**) is as follows:

1. Capture the biometric data from the appropriate sensor;

2. Extract the features from the captured image and stored it as a template;

3. The template of biometrics can be stored in smart cards, local machines or on a server for future use;

4. Scan the current biometric traits data;

5. For processing, from the image extract the features and from template;

6. For matching, take the input processed features with the existing biometric template;

7. On the basis of matching ranking score decide the business-level application and

8. Make the security evaluation of the system for proper use.

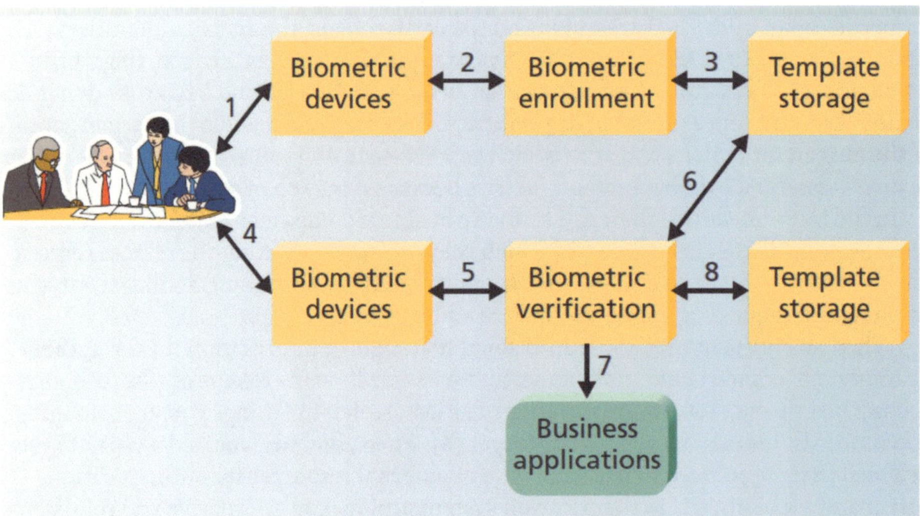

Figure 1.
How biometric system works.

2. Definitions of biometric

Biometrics is a way of automatic identification or verification of the person on their physical, chemical or behavioral properties. It is a scientific way to analyze the biological unique patterns of the individual person through the use of advanced technology. Biometrics is a scientific approach to understand and find the unique representation of the person. Biometrics is purely depending on the physical, behavioral or chemical properties of the human being for secure access or in identification and verification otherwise biometric devices have no use in authentication. Biometrics is a science of identification or verification of a person through the face, fingerprint or voice, etc. measurement of unique patterns. These unique patterns of the person called as features stored in embedded devices, smart cards are known as templates or bio-prints. They are used to verify the identity of the person by comparing them to the previously stored bio-prints [8].

3. Biometrics model

In general, the biometric model (**Figure 2**) is divided into five parts are as follows.

3.1 Data collection

The first part is a data collection which consists of biometric presentation and sensor. In this part, biometric modality is captured through the biometric sensor and it represents in its equivalent format for user understandable level. The biometric data sample is collected through various biometric traits either physical or behavioral. The biometric samples were taken from an instance, it should be unique at multiple impressions, iteration or frequent timely. At the time of data acquisition through the sensor, some technical issues may arise such as noise generated

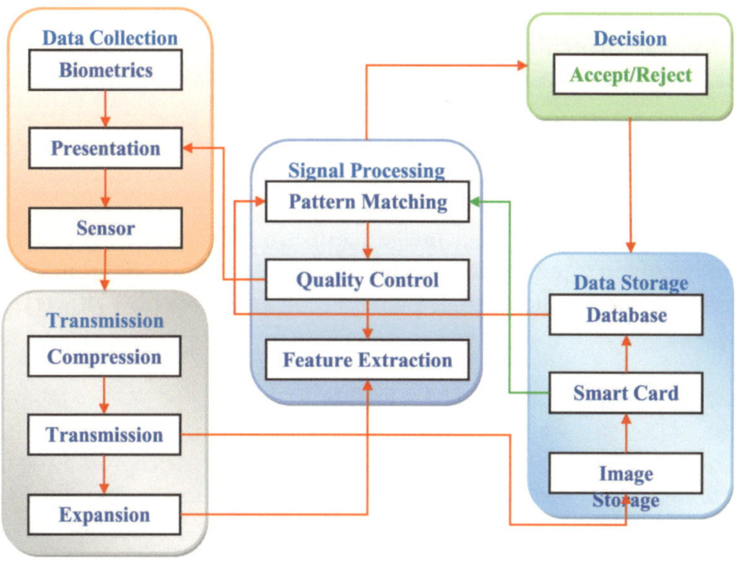

Figure 2.
The block diagram of biometric model.

in the background while taking the samples of speech or sensor sensing capacity fault. The user does not support while collecting the samples through the sensors. Sometimes more pressure is applied to the fingerprint device then noisy data will be captured.

3.2 Transmission and Data Storage

Sometimes at the time of storage, data are in large volume, we need to store it into the compressed format for fast transmission. At the time of compression technique, we need to be careful while selecting the algorithm otherwise there may be chances of adding more artifacts in original data samples.

It is not mandatory to store the data on the device, it might be stored on the local machine or the server as per the application requirement and cost-effectiveness. Sometimes, there is no need to store the data on the server or the application may be taken care of it to store it into the secure format on the same application device.

3.3 Signal processing

The main core component of any biometric system is signal processing, in which we can check the quality of the image, feature extraction or pattern matching. Sometimes due to distortion in input image, there is a chance of noisy image or bad quality data then there is a need to recapture the image or biometric samples once again. After ensuring the good quality data then proceed for the feature extraction through an appropriate technique that will be suitable for the application. Pattern matching is a key role player in which stored data template is matched with the given input samples. The pattern matcher will compare the matching results and send them to the decision module for the final decision.

3.4 Decision

After the pattern matching score, the decision module decides the acceptance or rejection of the person by using predefined certain threshold values [9].

4. Types of image acquisition devices

The camera is one of the famous image acquisition devices. Cameras are mainly divided into two main types i.e. analog and digital cameras. Digital cameras can be further classified into parallel digital, Camera Link and IEEE 1394 [10, 11].

5. Image acquisition process in MATLAB

MathWorks has developed a proprietary multi-paradigm programming language and numeric computing environment is known as Matrix Laboratory. From this matrix laboratory, MATLAB word is abbreviated. In MATLAB, we can perform matrix operations, plot the various graphs, develop the functions, interfaces and make the interfacing for the other programming languages programs.

MATLAB provides the programming and numeric computing platform for the analysis of data, algorithm development, creation of models, hence, it is widely used by scientists, engineers, researchers. If you wish to get more knowledge about the image acquisition process and capabilities (Image Acquisition Toolbox), MATLAB documentation is the best source.

Image Acquisition Toolbox (ImATool) provides the ability to handle the numeric calculation by using the predefined available functions. Under the IMAtool, wide functions are defined which supports the following image acquisition operations:

- Acquiring images through many types of image acquisition devices
- Acquiring images through many types of image acquisition devices
- From professional-grade frame grabbers to USB-based Webcams
- Viewing a preview of the live video stream
- Triggering acquisitions (includes external hardware triggers)
- Configuring callback functions that execute when certain events occur
- Bringing the image data into the MATLAB workspace

MATLAB has capabilities to extend the imtool in your own code or combination with other toolboxes, such as the Image Processing Toolbox and the Data Acquisition Toolbox. It also provides the Image Acquisition Blockset i.e. Simulink interface. This block set extends Simulink with a block that lets you bring live video data into a model. To get the live image data from the acquisition boards after plug-in for that Matlab provides the Data Acquisition Toolbox through which we can able to communicate with the acquisition boards.

For image processing, analysis and algorithm development related functions are defined under the Image Processing Toolbox. For control and communication with the test and measurement of various equipment's related functions are defined under the Instrument Control Toolbox. You can also perform the Video and Image Processing Blockset by using the Simulink model.

5.1 Basic image acquisition procedure

To develop a motion detection application, certain basic steps are required, which are shown in **Figure 3**. Pixel-to-Pixel variations in the scene show the difference in acquired image data frames in developed motion detection application. Sometimes frame will be constant, which means there is no change in incoming frame pixel values. Suppose, variation is found in the incoming image frame pixel values, it means, a change in the scene which is also capable to display in the application. Very few lines of coding are required for image frame data acquisition with the help of the toolbox, which is described in Section 5.2 examples. For the execution of the given code in the example, the image acquisition device should be connected to your system. Image acquisition devices can be professional so that the acquired image data frame will be quality image data for the high level of assumptions. Examples of professional devices

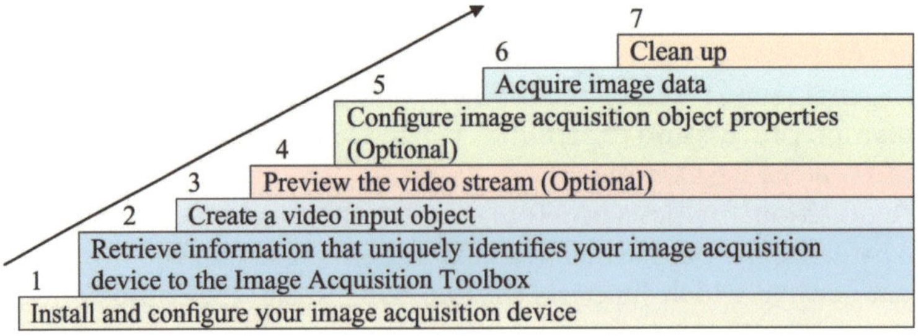

Figure 3.
Image acquisition basic steps.

are frame grabber, generic windows, webcam, etc. This sample code will be able to capture the image from different types of image acquisition devices by doing simply minor changes sometimes. **Figure 3** shows how to acquire image data with the help of Image Acquisition Toolbox and **Figure 4** shows Image Acquisition Toolbox Components.

5.2 Example: acquiring 10 seconds of image data

In this example, you can configure time-based acquisition using the number of frames per trigger.

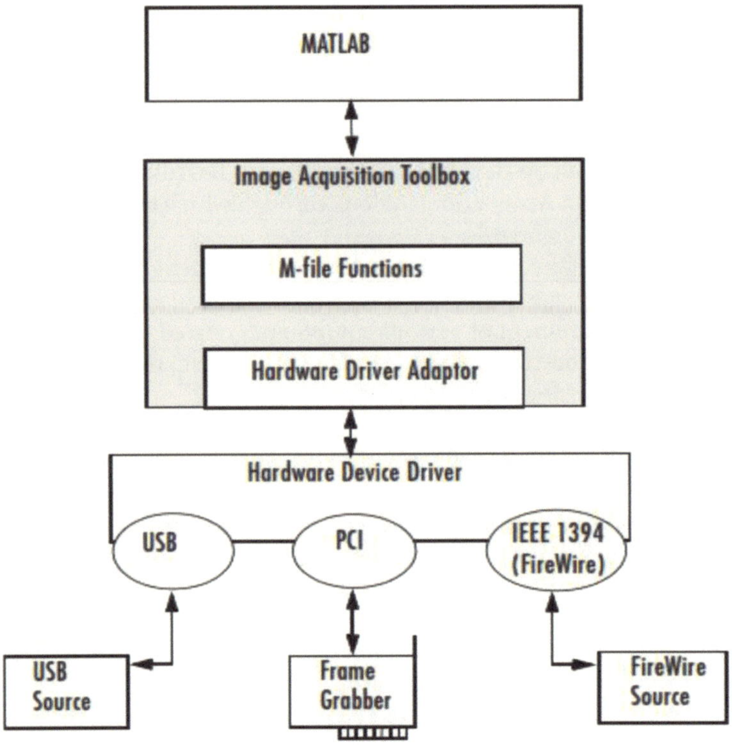

Figure 4.
Image acquisition toolbox components.

5.2.1 Create an image acquisition object

Before taking the input from the connected camera on your current system, you need to first create an object. The camera gives synchronous data, it is continuous information in the form of bits. To convert this information in the form of a visual display unit by using windowing techniques, therefore, it becomes video. From this video input, you want to capture the image. Hence, you need to create the video input object of your camera device for accessing the device property. You can check the list of image acquisition devices by using *imaqhwinfo* function. Also, you will get syntax and formats available for the respective devices in the form of structured data. By selecting the appropriate information from image acquisition devices, you can generate the windows-based output of your camera in the form of video. For the creation of video input object *videoinput()* function is available. You can pass two parameters while calling this function. The first parameter is the type of camera and the second is the camera ID. The syntax for the creation of camera object is:

vid = videoinput('winvideo',1);

In this case, the vid is the camera object, video input is the function, win video is a type of image acquisition device category and 1 is the camera id number.

5.2.2 Configure properties

Once the camera object has been created, you can acquire the image information at a specific time. If you wish to acquire the 10 or 20 seconds of data from your camera, then it has to be set the property as FramesPerTrigger. For the calculation of FramesPerTrigger first, check the frame rate of the camera per second and multiply it by the number of seconds. Then it can be considered for the camera configuration property.

Example. If the frame rate of the camera is 20 frames per second and you want to acquire the 10 seconds data then it will become 20 *10 = 200. To set this configuration there is set() function available in MATLAB. This function will receive three arguments: the first argument is video object i.e. vid, the second argument is configuration property i.e. FramesPerTrigger and the third one is the value of FramesPerTrigger i.e. as per example 200.

set(vid,'FramesPerTrigger', 200).

5.2.3 Start the image acquisition object

To acquire the image from the camera to our system, we have to start gabber of camera object in MATLAB as start() function available in MATLAB.

Ex. *start(vid).*

After calling the start() function video object is started and tries to store the temporary data into the memory buffer. It will acquire the image data continuously till the specific number of frames as in example 200. This process executes as a trigger when the start() function is called in our program and stops when the specific number of frames is received in the memory buffer.

Figure 5 shows the image preview when you start the video object.

Figure 5.
Image preview.

5.2.4 Bring the acquired data into the workspace

Load your data for verification that all data contains are accurate which we have planned to acquire the image as per our resolution and configuration. In MATLAB, there is a getdata() function that returns number of frames acquired within the specific time slot with a timestamp. We can verify the amount of acquired data according to timestamp and the difference between the first frame and the last frame.

Start Camera Code:

```
global vid;
vid = videoinput('winvideo',1);
vidRes = get(vid, 'VideoResolution');
nBands = get(vid, 'NumberOfBands');
set(gcf,'CurrentAxes',handles.axes1);
hImage = image(zeros(vidRes(2), vidRes(1), nBands));
preview(vid, hImage);
```

Once you have started the video object and set the bands, you can preview live camera acquisition data into the windows as shown in **Figure 6**. Then you can fix the face position into the camera preview and then follow the next steps to capture the preview image as shown in **Figure 7**.

Capture Image Code:

```
global vid;
%% Image Capture through the Current Video Preview.
global im;
im = getsnapshot(vid);
set(gcf,'CurrentAxes',handles.axes2);
imshow(im);
```

In this way, we have successfully captured images using MATLAB code. Now, we have to develop the face database for your face recognition application [12–15]. Then

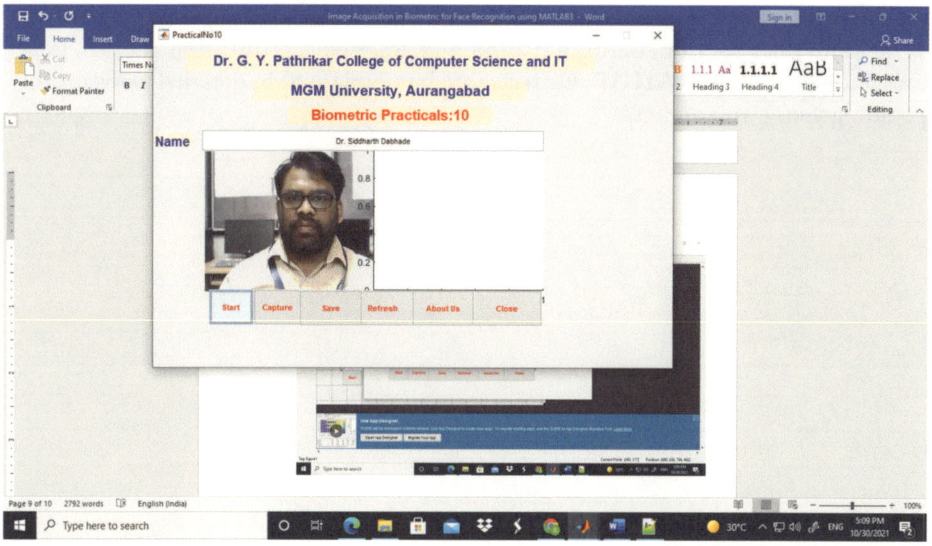

Figure 6.
Image preview in GUI.

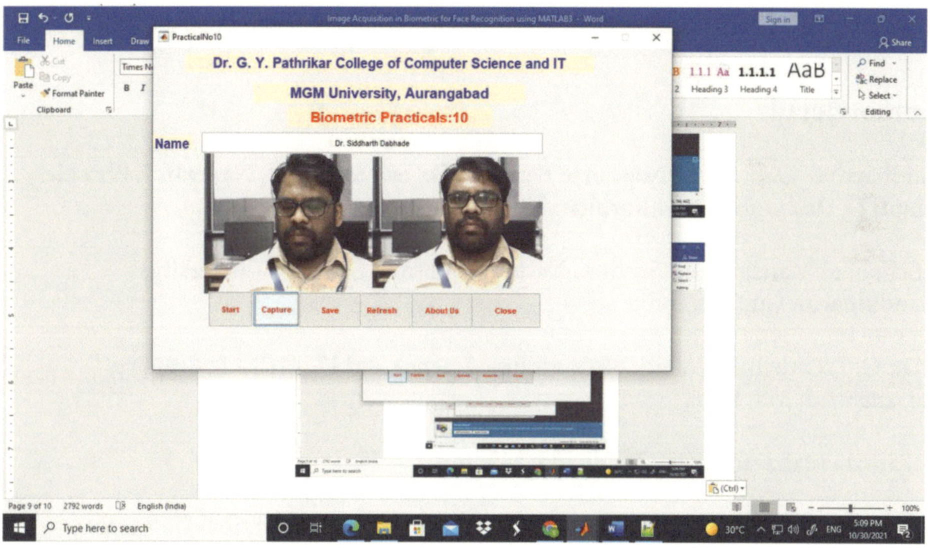

Figure 7.
Image preview in GUI after capture the image.

go for the feature extraction, classification and recognition level as per your preferred suitable techniques [16–18].

6. Conclusion

Biometrics is mostly used for authentication purposes in security. Face recognition in real-time itself is a challenging task. For that taking input from the camera is very important for real-time application. In this chapter, we have mainly focused on how to acquire the face image using MATLAB. The complete chapter is divided into five sections introduction, definition of biometrics, image acquisition devices, image acquisition process in MATLAB. Each section has explained in detailed steps for the upcoming young researchers.

Author details

Siddharth B. Dabhade[1*], Nagsen S. Bansod[2], Yogesh S. Rode[3*], Narayan P. Bhosale[4], Prapti D. Deshmukh[2] and Karbhari V. Kale[5]

1 School of Management Studies, National Forensic Sciences University, Gandhinagar, Gujarat, India

2 Dr. G. Y. Pathrikar College of Computer Science and IT, MGM University, Aurangabad, MS, India

3 Jijamata Mahavidyalaya, Buldhana, MS, India

4 Department of Computer Science, Indira Gandhi National Tribal University, Amarkantak, Madhya Pradesh, India

5 UDCSIT, Dr. Babasaheb Ambedkar Marathwada University, Aurangabad, MS, India

*Address all correspondence to: dabhade.siddharth@gmail.com
and ys.rode@gmail.com

References

[1] Jain AK, Ross AA, Nandakumar K. Introduction to Biometrics. Vol. XVI. 312. Boston, MA: Springer; 2011. p. 196. DOI: 10.1007/978-0-387-77326-1

[2] Jain AK, Patrick F, Ross AA, editors. Handbook of Biometrics. Vol. X, 556. Boston, MA: Springer; 2008. p. 60. DOI: 10.1007/978-0-387-71041-9

[3] Adeoye OS. A survey of emerging biometric technologies. International Journal of Computer Applications (0975-8887). 2010;**9**(10):1

[4] Tripathi KP. A comparative study of biometric technologies with reference to human interface. International Journal of Computer Applications (0975-8887). 2011;**14**(5):10

[5] Dhir V et al. Biometric Recognition: A Modern Era for Security. International Journal of Engineering Science and Technology. 2010;**2**(8):3364-3380

[6] Liu S, Sullivan J, Ormaner J. "A practical approach to enterprise IT security," in IT Professional. Vol. 3, No. 5. Manhattan, New York, U.S.: IEEE; 2001. pp. 35-42. DOI: 10.1109/6294.952979

[7] Rosenzweig P, Kochems A, Schwartz A. Biometric Technologies: Security, Legal, and Policy Implications. Vol. 12. Washington, D.C., U.S.: Legal Memorandum, The Heritage Foundation; 2004. Available from: https://www.heritage.org/homeland-security/report/biometric-technologies-security-legal-and-policy-implications#

[8] Dahiya N, Kant C. Biometrics Security Concerns. In: 2012 Second International Conference on Advanced Computing & Communication Technologies, Rohtak, India. Manhattan, New York, U.S.: IEEE; 2012. pp. 297-302. DOI: 10.1109/ACCT.2012.36.

[9] Tiwari S, Zhai G, Carter SA, Tiwari S. Evaluating the capability of biometric technology. International Journal of Advanced Research in Computer Engineering & Technology. 2012;**1**(2):18

[10] Image Acquisition, White Paper, Available from: https://www.ni.com/en-in/innovations/white-papers/06/image-acquisition.html Accessed: June 21, 2021

[11] Nudelman S. Image Acquisition Devices and Their Application to Diagnostic Medicine. In: Höhne KH, editor. Pictorial Information Systems in Medicine. NATO ASI Series (Series F: Computer and Systems Sciences). Vol. 19. Berlin, Heidelberg: Springer; 1986. DOI: 10.1007/978-3-642-82384-8_2

[12] Kazi MM, Rode YS, Dabhade SB, Al-Dawla NNH, Mane AV, Manza RR, et al. Multimodal Biometric System Using Face and Signature: A Score Level Fusion Approach. Advances in Computational Research. 2012;**4**(1):99-103

[13] Dabhade SB, Bansod N, Kazi MM, Rode YS, Kale KV. Hyper Spectral Face Recognition Using KPCA. International Journal of Scientific & Engineering Research (IJSER). 2016;7(10):548-550

[14] Dabhade S, Bansod N, Kazi M, Rode Y, Kale K. Hyper spectral face recognition using Gabor + KPCA. IOSR Journal of Computer Engineering (IOSR-JCE). 2017;**2**:61-64. Available from: https://www.iosrjournals.org/iosr-jce/papers/Conf.17003/Volume-2/12.%2061-64.pdf

[15] Siddharth BD, Bansod NS, Rode YS, Mkazi M, Kale KV. Performance Evaluation on KVKR- Face Database using Multi Algorithmic Multi Sensor Approach. International Journal of Computer Applications. 2018;**180**(13):30-36

[16] Dabhade SB, Bansod NS, Rode YS, Kazi MM, Kale KV. Multi sensor, multi algorithm based face recognition & performance evaluation. Vol. 21-23. Jalgaon, India: IEEE, 2016, International Conference on Global Trends in Signal Processing, Information Computing and Communication (ICGTSPICC); 2016. pp. 113-118. DOI: 10.1109/ICGTSPICC.2016.7955280

[17] Siddharth BD, Bansod N, Naveena M, Khobragade K, Rode YS, Kazi MM, et al. Double Layer PCA based Hyper Spectral Face Recognition using KNN Classifier. Mysuru, Karnataka, India: IEEE, International Conference on Current Trends in Computer, Electrical, Electronics and Communication (ICCTCEEC), Vidyavardhaka College of Engineering; 2017. pp. 119-122

[18] Dabhade SB, Rode YS, Kazi MM, Manza RR, Kale KV. Face Recognition using Principle Component Analysis and Linear Discriminant Analysis Comparative Study. Kurukshetra, Haryana: Elsevier, 2nd National Conference on Advancements in the Multi-Disciplinary Systems, Technology Education & Research Integrated Institutions (TERII); 2013. pp. 196-202 [8 Citations]

Chapter 10

Your Vital Signs as Your Password?

Hind Alrubaish and Nazar Saqib

Abstract

Cognitive biometrics (vital signs) indicate the individual's authentication using his/her mental and emotional status specifically, electrocardiogram (ECG) and electroencephalogram (EEG). The motivation behind cognitive biometrics is their uniqueness, their absolute universality in each living individual, and their resistance toward spoofing and replaying attacks in addition to their indication of life. This chapter investigates the ability to use the vital sign as unimodal authentication in its status by surveying the recent techniques, their requirements and limitation, and whether it is ready to be used in the real market or not. Our observations state—that the vital signs can be considered as a PASSWORD due to their uniqueness, but it needs more improvements to be deployed to the market.

Keywords: electrocardiogram, ECG, electroencephalogram, EEG, electrooculography, EOG, blood flow, vital sign, authentication, recognition, biometrics

1. Introduction

Our mobiles, laptops, houses, and cars, rely on identification and authentication procedures to protect ourselves, data, and assets. Different methods are existing for this purpose which differ in their way and security level. These methods were ranging from traditional techniques where the user must "know" or "have" such as passwords, keys, or cards, to biometric techniques that define the user himself. Scientists tried in the last two decades to focus on biometric techniques to avoid problems associated with traditional ones, such as loss, theft, forgery, or coping. Biometric techniques defined the individual's characteristics and required his/her physical presence to access the system without the need to carry or memorize anything. Unlike the traditional techniques, biometrics cannot be shared with anyone.

To identify any feature as a biometric, the following requirements should exist; *Universality* where each person should have this feature, *Distinctiveness* where the feature should uniquely identify each person, *Collectability* where the feature can be measured quantitatively, *Performance* where the feature can be measured in term of its accuracy, time, error rate ... etc., *Acceptability* where the user can accept to use the feature as an authentication technique, *Circumvention* showing how easy the user will bypass the system [1].

Many human features achieved these requirements and are labeled as biometric techniques where it can be categorized into; behavioral, physiological, and cognitive. Behavioral biometrics deal with functional features, such as voice, gait, signature, and keystroke. Physiological biometrics deal with anatomical features, such as fingerprint, face, iris, and ear shape. Cognitive biometrics use a biological signal generated from the heart, brain, or automatic nervous system which is an indicator of the individual's mental and emotional states, such as electrocardiogram (ECG) and electroencephalogram (EEG).

Cognitive biometrics outweigh behavioral and physiological biometrics as it cannot be acquired, falsified, manipulated, or copied by external attackers [2] another advantage it can be utilized as a liveness detector.

This chapter reviews the state-of-the-art of human vital signs (cognitive biometrics) as biometric authentication. It will involve the recently discovered techniques, their description, limitation, and applications. This chapter is organized as follows: Section two investigates the electrocardiogram (ECG), while section three investigates electroencephalogram (EEG). Section four describes electrooculography as an authentication technique. Section five cites the blood flow as a patent to be used as a biometric. While section six discusses the ability of the vital signs to be used as unimodal authentication. Finally, section seven concludes this chapter.

2. (Heart-Beat Print) using electrocardiogram (ECG)

Electrocardiogram (ECG) is a recording of the electrical activity produced by the heart by placing electrodes on the body's skin to obtain the signals originating from the heart muscle. Any ECG consists of three components; P waves represent atria contractions (left and right), QRS reflect ventricular contractions (left and right) and appeared as a series of three waves, and T wave represents the electrical activity produced by the ventricular when it charging for the next contraction (repolarization), each ECG signal has six peaks and valleys [3–6]. Individual's ECG varies from one person to another based on the physiological, anatomical, and geometrical conditions, in addition to the position and size of the heart, also age and sex play a role in its uniqueness. Therefore, it can be used as an authentication technique [4].

Every living person can produce ECG therefore, the universality requirement is satisfied. Moreover, it is a proof of life which means that the ECG is more universal than any other physiological and behavioral biometrics. The extracted features vary for each person where the distinctiveness requirement has been achieved. These features can be measured quantitively using a standard available system which proves its collectability requirement. Although these systems are already in use for the patient within the medical field but not widely accepted in daily use. Finally, circumvention is achieved as we can measure how much easy the intruder will bypass the ECG authentication system. This is more difficult than the other biometric features as the ECG cannot be falsified or manipulated and require a living individual to authenticate his identity. As a result, the ECG can be considered a biometric authentication.

Any ECG-based authentication system comprises the following steps:
(1) Acquisition: Electrodes placed on the body's skin to capture the signals.
(2) Quality Assessment: The system preprocesses the captured data to eliminate the noise and appropriately represent the signal. (3) Feature Extraction: The system extracted and normalized the features in two approaches; Fiducial Approach: The system detects, process, and classify the three waves P, QRS, and T based on their peaks, boundaries, and intervals between them. Non-Fiducial Approach: The system applies time or frequency analysis to obtain statistical features [7]. (4) Finally, Decision: The system classifies the extracted features to make the authentication decision [5, 8].

Numerous studies deliberate how the ECG is effective as a biometric, the following studies illustrate different approaches and algorithms.

In ref. [9], the authors proposed an identification technique based on ECG and musical features. After pre-processing ECG recordings, they transform them into audio wave files, split them into segments, and extract five musical dimensions to be faded into the classifier. They used MIT-BIH Normal Sinus Rhythm dataset. The proposed technique achieved 96.6% accuracy.

In ref. [10], the authors proposed EDITH, a deep learning-based framework for ECG Biometric Authentication systems. They demonstrate that Siamese architecture can be used over typical distance metrics to improve performance. They evaluated EDITH in four datasets using a single heartbeat. Their accuracy reached (96–99.75%) which can be enhanced using multiple heartbeats. The proposed framework reduced the Equal Error Rate to 1.29%.

In ref. [11], authors proposed two Model CNN and RestNet-Attention using ECG Signals where the signals are authenticated using an end-to-end structure without any handcrafting preprocessing, feature extraction, and classification which reduced the computational complexity. The proposed algorithm achieved 98.59 and 99.72% accuracy using PTB and CYBHi datasets.

To address the individuality issue of ECG over a larger population, authors in ref. [12] the present non-fiducial approach of ECG authentication and identification. They used autocorrelation and a combination of three transformation techniques DCT, DFT, and WHT to extract the features. Then the performance of these techniques has been evaluated on two-dimensionality reduction techniques—PCA and LDA. The best accuracy results achieved using DFT and LDA in QT Database (100%).

In ref. [13], the authors proposed a Dynamic Time Wrapping (DTW) algorithm to provide identification and authentication to the authorized people using ECG signals in Wireless Medical Devices (WMD). They used DTW to measure the correlation between different ECG records. They used Physionet dataset that contains 20 subjects of all ages with 310 records including abnormal ECGs, and a long period interval between ECG recordings to increase the reliability. They achieved a 99.9% accuracy rate.

In ref. [14], the authors proposed an algorithm to authenticate users with their doctors remotely using ECG signals. The algorithm consists of two parts; a registration process where the Discrete Wavelet Transform (DWT) extracts the features to be stored. The second part is the authentication process where the features will be matched with existing templates using the Sum of Squared Differences (SSD). They utilized the ECG IDDB Physionet dataset, and one lead has been used to fit in IoT devices, the algorithm uses non-Fiducial features, and achieved 91% accuracy.

In ref. [15], the authors develop an authentication algorithm using Linear Discrimination Analysis (LDA) to classify 16 subjects taken from the Physionet dataset based on their ECG signals (each one has 75–150 heartbeats); they extracted eight fiducial features from the ECG where they achieved 92.69% accuracy rate. The algorithm is scalable to large databases.

In ref. [16], the authors introduced authentication technology to record ECG signals of 55 voluntary subjects before and after insensitive exercise for five minutes using two positions; rest and sitting. LDA was used for feature extraction and classification. The best accuracy achieved within five minutes of recording is 96.11%.

In ref. [17], the authors proposed a framework for authentication using ECG where they used a Neural Network (NN) as a classifier. The test was not successful considering the small size of the dataset.

In ref. [18], the authors apply four nonlinear methods to extract fiducial features for the ECG authentication system; Generalized Hurst Exponent (GHE), Detrended Fluctuation Analysis (DFA), Higuchi's Fractal Dimension (HFD), and Rescaled Range Analysis (RSA). A record of 18 subjects from the MIT-BIH Normal Sinus Rhythm Database fed into SVM as a classifier that achieved a 99.06% accuracy rate. The results showed that GHE has the optimal index to authenticate the subjects.

In ref. [19], the authors propose the use of long short-term memory (LSTM)-based Recurrent Neural Networks (RNN) to use ECG as an authentication solution where there is no feature extraction. The method has been applied to ECG-ID and MIT-BIH Arrhythmia (MITDB) datasets. They achieved a 100% accuracy rate. As the number of subjects increases, the equal error rate drops.

In ref. [20], the authors proposed a method using phase-space reconstruction (PSR) of a single lead of ECG. They used a time delay technique to reconstruct the ECG's signal into phase space to find the best identifiable time-delay value. Twenty-one geometric features have been extracted in different situations: rest, during exercises, listening to music, and watching a movie. The procedure was conducted on 31 subjects and the accuracy rate was 97.7% when the delay is 8 ms.

In ref. [21], the authors proposed an identification method by extracting five fiducial points using Empirical Mode Decomposition (EMD). Hidden Markov Model (HMM) has been used as a classifier with the Bakis model on 44 subjects from the MIT-BIH Arrhythmia database. The method achieved a 98.52% accuracy rate.

In ref. [22], the authors proposed a mobile authentication algorithm based on ECG where the user will need to touch only two electrodes (lead I) of the mobile device to be authenticated. The experiments were conducted on ten subjects in addition to 37 records from the Physionet dataset. The algorithm uses a hierarchal scheme that reduces the acquisition time to 4s.

The following table summarizes the previous studies to use ECG as biometric authentication (**Table 1**).

Moreover, different scientists propose various utilization of the ECG besides authentication. In ref. [23], researchers at Binghamton University developed a robust and reusable authentication and data encryption means to protect the patients' health records using their heartbeat (ECG) where the cost, time, storage, and complexity will be much more effective than using traditional encryption solutions. In ref. [24], the authors use ECG steganography to secure patient's confidential information. Another use is generating a secret key for data encryption and enhanced security in personal wearable devices using a patient's ECG [25]. In ref. [26], the authors proposed software for remote interaction between the cardiovascular disease patients and the health provider to monitor their ECG, blood pressure, and heart rate.

As a summary of the previous studies, we can observe that there is some limitation that may lessen the ECG's effectiveness as a biometric which needs further studies to be addressed. (1) The performance of ECG depends on how (P, T, QRS) are detected accurately. (2) Heart Rate Variability: Many factors can affect the ECG morphology which can be classified into short-term and long-term factors. In the short term where physical activity, mental status, drinking caffeine ... etc. can affect the ECG, while the long-term factors are the change in the lifestyle such as using the medication, or heart diseases [4]. (3) The size of tested subjects does not exceed 300 which indicates ECG has not proven its ability within a large population to be deployed to the market as an authentication technique; more studies need to be done to confirm its scalability. (4) Also, there is no one study has studied the issue in the case of the heart transplant and whether it will affect the ECG authentication process or not. (5) Similarly, in the

DOI: http://dx.doi.org/10.5772/intechopen.104783

Reference #	Publication Year	Purpose	Method	Database	# of Subjects	# of Records	Approach	Accuracy
[9]	2022	Identification	Deep Learning (Transform ECG records into sound wave files characterized with musical features for human identification)	MIT-BIH	18	—	non-fiducial	96.6 %
[10]	2021	Authentication	EDITH, a deep learning-based framework	ECG-ID, MIT-BIH Arrhythmia – PTB Diagnostic ECG Database – MIT-BIH NSRDB	90-47-290-18	—	—	96.247%-98.170%-99.702%-99.500% (closed environment)
[11]	2020	Authentication	two end-to-end deep neural networks (CNN and ResNet)	PTB and CYBHi	290 - 65	—	—	98.85 and 99.27%
[12]	2019	Identification & Authentication	Autocorrelation (AC) with DCT -DFT-WHT Then Then PCA & LDA	MIT-BIH arrhythmia & QT database	48-39		non-fiducial	DFT & LDA (99.98% (99.83%) for QT DB
[13]	2018	Authentication	DTW	Physionet ECG-ID	20	310 ECG	—	99.9%
[14]	2017	Authentication for Remote patient in IoT	Template Matching SSD	Physionet - IDDB dataset	90	N/A	Non-Fiducial	91%
[15]	2017	Authentication	LDA	Physionet	16	5 or more for each subject	Fiducial	92.69%

Reference #	Publication Year	Purpose	Method	Database	# of Subjects	# of Records	Approach	Accuracy
[16]	2017	Authentication with the ECG data recorded after the harsh exercise	LDA	University of Toronto Database (UofTDB)	55	N/A	Fiducial	The subject recognition accuracy was 59.64%, which is too low to utilize, after one minute the accuracy was higher than 90% and it increased up to 96.22% within 5 minutes, which is plausible to use in authentication circumstances
[17]	2017	Authentication	RF	Physikalisch-Technische Bundesanstalt (PTB) Diagnostic ECG Database	290	549	-	88.45%
[18]	2017	Authentication	SVM	MIT-BIH Normal Sinus Rhythm Database	18	N/A	Fiducial	99.06
[19]	2017	Authentication	LSTM-based RNN	ECG-ID & MIT-BIH Arrhythmia (MITDB)	90 47	310	No extraction	100%
[20]	2017	Authentication	SVM	Voluntarily subjects	13	—	Fiducial	97.7%
[21]	2016	Authentication	Hidden Markov model (HMM) classifier with Bakis model	MIT-BIH Arrhythmia (MITDB)	44	—	Fiducial	98.52%
[22]	2016	Mobile Authentication	—	Voluntary Subjects	10		Fiducial	—

Table 1.
A comparison of the latest studies in ECG.

case of the twins, whether there is any matching that can breach the confidentiality of the authentication process?

3. Brain prints using electroencephalogram (EEG)

Electroencephalogram (EEG) signals are the representation of the brain activity in the neurons either in the baseline task (relaxed) situation or in response to a functional status such as sleeping, solving a mathematical problem, reading some text, or having some diseases. These activities generated signals captured by placing electrodes on the scalp. There are five different waves in each EEG; Alpha wave appears during relaxation. Theta waves appear in the quite focus, short-memory tasks, and memory retrieval. Beta waves appear in a normal working rhythm; such as increased alertness, and anxious thinking. Delta wave happened during deep sleep. Gamma waves represent active information processing [27, 28]. Unlike other biometric techniques, the user can change the password by changing the mental task itself. EEG cannot be copied since it represents the real status of the brain.

As the ECG, every living person can produce EEC therefore, the universality requirement is satisfied also, it is aliveness detection. Each EEG has a different pattern in terms of its wave shape where the distinctiveness requirement has been achieved. These features can be measured quantitively using portable devices which proves its collectability requirement. The Acceptance of EEG may it will be a quite little difficult among the users, to raise the level of acceptance of EEG among the users the following may be done—(1) the typical EEG device consists of a number of electrodes that may be needed to be minimized into three or four [29]. (2) The use of dry electrodes instead of wet ones. Finally, the circumvention of EEG cannot be occurring as the spoofing in EEG is not possible. In addition to that, any intruder will not be able to generate a real EEG and impersonate the real user.

For each EEG-based authentication system, the following steps must occur; (1) Acquisition: EEG is captured using electrodes placed over the scalp where the subject is exposed to a specific task. Each electrode collects a wave for a specific region within the brain where all the waves will be combined into one. (2) Quality Assessment: The system preprocesses the captured signals to eliminate the noise and represent the signal in an appropriate way. (3) Feature Extraction: The system extracts and normalized the features. (4) Finally, Decision: The system classifies the extracted features to make the authentication decision [5, 8].

Numerous studies deliberate how the EEG is effective as a biometric, the following studies illustrate different approaches and algorithms.

In ref. [30], the authors proposed MusicID, a behavioral biometric framework for smart headset-enabled IoT environments. MusicID is induced by the user's brain's response to two forms of music: Common English songs and an individual's favorite song. Their analysis showed that Alpha and Beta waves have more predictive capabilities. The framework achieved 98% for user identification and 97% for user verification.

In ref. [31], the authors designed electroencephalogram authentication access control for the smart car. The accuracy results achieved 87.3%

In ref. [32], the authors proposed an ECG authentication system using neurological responses to music. They used Alpha and Beta waves collected from seven electrodes. KNN is used to classify the data. They achieved 76.4%–92.3% accuracy results.

In ref. [33], the authors proposed a method to denoise the ECG signals based on the multi-objective Flower Pollination Algorithm and Wavelet Transform to extract the features. The test was conducted using an EEG motor movement/imagery dataset.

In ref. [34], the authors used power spectral density analysis to analyze EEG signals which fed into KNN to classify the EEG. The achieved accuracy was 89.21%.

In ref. [35], the authors proposed a pragmatic authentication system using EEG. They collected EEG of 29 subjects using a single dry electrode via a cheap Neurosky Mindwave headset and ten subjects using 14 electrodes via Emotive. The achieved accuracy for the first group was 80% while the second group achieved 92.88%.

In ref. [36], the authors studied how the differences in the emotional states affect the classification performance. The results showed that there is better performance when the subjects have the same emotional status.

In ref. [37], the authors proposed a biometric system using an in-ear EEG sensor where there is no need for skilled assistance or preparation. The results showed equivalent results to the on-scalp recording.

In ref. [38], the authors proposed an authentication framework using self or non-self-face images which were applied using Rapid Serial Visual Presentation (RSVP).

In ref. [39], the authors proposed an identification framework to identify users while they are listening to four genres of music.

In terms of the band type's performance, authors in ref. [40] present a superior performance of power spectral density features of gamma band during the rest state over the delta, theta, alpha, and beta of EEG signals.

In ref. [41] investigate the most effective frequency bands for authentication purposes using EEG signals at the rest status via Neural Networks (NN) as a classifier. The results show that beta has the best performance while delta gave the worst performance.

Another study [42] found that extracted feature from the gamma band in the left-posterior quarter of the brain has more reliable and stable information regardless of the emotional status. They classify the signals using five features and SVM as a classifier.

The following table summarizes the previous studies to use EEG as biometric authentication (**Table 2**).

Moreover, the authors in ref. [43], presented a monitoring and safety platform consisting of automotive sensors to capture real-time information about the driver and the vehicle in addition to a wearable body sensor network to collect the driver's EEG and ECG. They investigate the effect of the driver's behavior on road conditions. The experiment was conducted on five subjects via 16 dry electrodes using theta and beta bands. The results showed that these biometrics could be used detection of driver distortion.

From the previous studies and as well as the ECG, EEG has its limitations that need to address to raise the effectiveness of the EEG as a unimodal authentication system; (1) the acquisition process is quite difficult as the electrode cap needs a significant effort to place it above the head in specific places. Most of the used acquisition equipment was a medical cap, and it needs to be simplified. (2) Different factors may affect the EEG, such as stress and general arousal. Therefore, it may not authenticate the right person. (3) EEG acquisition has a low power signal which needs a controlled environment. (4) the size of tested subjects does not exceed 150 which indicates ECG has not proven its ability within a large population to be deployed to the market as an authentication technique; more studies need to be done to confirm its scalability. (5) similarly, to ECG, in the case of the twins, is there any matching that can breach the confidentiality of the authentication process?

Reference #	Publication Year	Purpose	Techniques	Database	# of Electrodes	Task	# of Subjects	Accuracy %	Band Type
[30]	2021	Authentication for IoT	Random Forest classifiers	Real Users	4 electrodes	Listening to Music	20	98% Accuracy for user identification and 97% accuracy for user verification	Alpha, Beta, Theta, Delta, Gamma, and raw EEG
[31]	2020	Authentication access control to smart car	Fisher distance analysis method	Real Users	40-channel neuroscan amplifier was used to collect EEG signals	Imagery Tasks	10	87.3%	—
[32]	2019	Authentication	KNN	—	7	Listening to Music	—	76.4% - 92.3%	Alpha, Beta
[33]	2018	Authentication	NN	EEG motor movement/ imagery	64 electrodes	Several motor/ imagery tasks	109	—	—
[34]	2018	Authentication	KNN	—	—	Visualization	—	80% - 89.21%	Combined theta, alpha, beta, and gamma
[35]	2018	Authentication	SVM - RLR - LDA	Mindwave	Single dry electrodes	—	29	80%	—
				Emotiv Epoc+	14 electrodes				

Reference #	Publication Year	Purpose	Techniques	Database	# of Electrodes	Task	# of Subjects	Accuracy %	Band Type
[36]	2018	Identification	LSVM - RSVM – KNN - MLP – (AdaBoost with DT)	DEAP, MAHNOB-HCI, SEED		—	32 (DEAP) 27 (MAHNOB-HCI) 15 (SEED)	99.51 (KNN) in (DEAP) 95.89 (LSVM) in (MAHNOB-HCI) 94.75% (LSVM) in (SEED)	—
[37]	2018	Authentication	Cosine Distance, SVM, LDA,	Two in-house datasets	In-ear sensor with two electrodes	Resting State	15, 5	95.7%	Alpha
[38]	2018	Authentication	HDCA	In-house	16	Visualization	45	91.46%	-
[39]	2017	Authentication	HMM, SVM	In-house	-	Listening Music (devotional, electronic, classical and rock)	60	97.50 % (HMM) 93.83 % (SVM)	Gamma, Beta, Alpha, Theta, Delta
[40]	2017	Authentication	-	PhysioNet	—	Rest State	109	0.001 (64 channels) 0.002 (19 channels)	(PSD) features of gamma band
[41]	2017	Authentication	NN	In-house	—	Eyes closed and solving some specific mathematical problem mentally	3	Beta (98.20% - 100%) Delta (92.82%-95.67%)	All

Reference #	Publication Year	Purpose	Techniques	Database	# of Electrodes	Task	# of Subjects	Accuracy %	Band Type
[42]	2017	Authentication	SVM	DEAP	—	1) mixture of emotional states; 2) the same specific emotional states; 3) different emotional states	32	88% - 99%	Gamma

Table 2.
A comparison of the latest studies in EEG.

For both ECG and EEG, we cannot guarantee that the user will generate the same signals under different factors such as mental status, age, etc. We may be able to eliminate this issue by registering the user in a periodic way under different situations [28].

4. Eye blinking waveform using electrooculography

Electrooculography (EOG) signals are the representation of generated signals due to eyeball or eyelid movements. These signals are generated once the eyeball rotates from its axis, and it is detectable by the electrodes placed around the eye. A positive deflection is generated in the signal when the eyeball rotates upwards or the eyelid closes and a negative deflection is generated when the eyeball rotated downwards or the eyelid opens [44].

These movements are captured by placing electrodes placed around the eye. There are five different waves in each EEG; the Alpha wave appears during relaxation. Theta waves appear in the quite focus, short-memory tasks, and memory retrieval. Beta waves appear in a normal working rhythm. Such as increased alertness, and anxious thinking. Delta wave happened during deep sleep. Gamma waves represent active information processing [27, 28]. Unlike other biometric techniques, the user can change the password by changing the mental task itself. EEG cannot be copied since it represents the real status of the brain.

In ref. [44], the authors adopt human recognition eye blinking where a preprocessing stage has been conducted to isolate EOG signals from EEG signals. They used time delineation as a discriminative feature. The experiment was done using the Neurosky Mindwave headset, which is used mainly for EEG signals, but the sensor arm can be used for this purpose.

5. Blood flow

A patent has been published in 2018 by SAMSUNG ELECTRONICS CO titled "Real Time Authentication Based on Blood Flow Parameters," the patent declared that we could use the blood flow as an authentication technique using a wearable sensor. The sensor detects at least the first physiological biomarker of the blood and the first morphological characteristic of the blood to determine the individual's uniqueness [45]. So far, no studies have explored and dealt with this patent.

6. Limitations

Despite the limitation of the vital signs as an authentication technique, there are promising features that can outweigh, and overcome the limitations. Vital signs characterize by their confidentiality and resistance to the spoofing attack as it is corresponding to emotional or mental status moreover, the users cannot authenticate themselves unwillingly as it will generate different signal statuses. Therefore, the Identity cannot be impersonated, copied, or captured from a distance. Also, it is impossible for an intruder to force the user to authenticate as it is subject to his mental status in some situations not under stress [28]. And most importantly, the vital sign is a liveness detector as it needs a live person recording. Unlike the face, finger, and eyes, the brain and heart have a rare chance to be injured.

However, it can be used as a multi-authentication system, a continuous authentication, or unimodal in specific cases until all the issues will be eliminated. Several domains can utilize the EEG and ECG signals in their current status. In the following we have proposed some applications to use ECG and EEG biometrics:

Anti-ATM Theft Model:

ECG sensors can be placed in the ATM to authenticate users using their ECG, which requires a previous registration of different user's emotional status (e.g., rest, horrified). The approach will be effective when the user is under attack from a burglar to withdraw an amount of money. The system will detect if the user is in an abnormal condition (horrified under coercion), and it will block the transaction.

Anti-Car Theft Model:

The proposed model will be based on ref. [43] where it can detect whether the driver is in a distraction mode or not in addition to that, it will prevent stealing the car or using it to commit a crime. The model can take advantage of either ECG or EEG as biometric authentication, ECG's sensors can be placed on the steering wheel, while the EEG can be placed in front of the headrest and behind the driver's head.

Top Secure Entities:

EEG and ECG can be used in sensitive and top secure entities, such as military and nuclear power reactors even in their status as they cannot be spoofed at all. A lair detector will be combined with the system and utilized the EEG and ECG to authenticate and verify the reason behind the access.

Continuous Authentication:

EEG and ECG can be used as a way for continuous authentication, such as the remote interaction in the online games to authenticate that the real user is who is claiming during the session game. The implementation of EEG and ECG within the online game environment can be accepted as the player wearing the headset and holding the control in his hands all the time.

7. Conclusion

This chapter surveyed the work done within the field of cognitive biometric authentication (vital signs) in terms of its limitation, requirement, advantages, and disadvantages specifically the ECG and EEG signals. Moreover, it investigated and raised some issues within the field that have not been studied yet and need to be addressed. Also, a recent patent on blood flow and electrooculography has been cited which can be considered a biometric authentication within the vital signs.

As an answer to our question, Can Your Vital Signs be Your Passwords? Yes, we can make sure that the heartbeat, brain waves, eye blinking, and blood flow act as a PASSWORD, but it cannot be used as a unimodal authentication approach in its current shape until their issues will be eliminated.

Conflict of interest

The authors declare no conflict of interest.

Author details

Hind Alrubaish* and Nazar Saqib
College of Computer Science and Information Technology, Imam AbdulRahman Bin Faisal University, Dammam, Saudi Arabia

*Address all correspondence to: haalrubaish@iau.edu.sa

References

[1] Dasgupta D, Roy A, Nag A. Advances in User Authentication. Springer; 2017

[2] Shdefat AY, Il Joo M, Choi SH, Kim HC. Utilizing ECG waveform features as new biometric authentication method. International Journal of Electrical Computer Engineering. 2018;**81**(2):658-665

[3] Electrocardiogram (ECG), 018. [Online]. Available from: https://www.nhs.uk/conditions/electrocardiogram/ [Accessed: 16 October 2018]

[4] Odinaka I et al. ECG biometric recognition: A comparative analysis. IEEE Transactions on Information Forensics and Security. 2012;**7**(6):1812-1824

[5] Safie SI, Soraghan JJ, Petropoulakis L. ECG biometric authentication using Pulse Active Width (PAW). In: 2011 IEEE Workshop on Biometric Measurements and Systems for Security and Medical Applications (BIOMS). 2011. pp. 1-6

[6] Keshavamurthy TG, Eshwarappa MN. Review paper on denoising of ECG signal. In: Proceedings of the 2017 2nd IEEE International Conference on Electrical, Computer and Communication Technologies, ICECCT. 2017

[7] Karimian N, Woodard DL, Forte D. On the vulnerability of ECG verification to online presentation attacks. In: 2017 IEEE International Joint Conference on Biometrics (IJCB). 2018. pp. 143-151

[8] Ribeiro Pinto J, Cardoso JS, Lourenco A. Evolution, current challenges, and future possibilities in ECG Biometrics. IEEE Access. 2018;**6**:34746-34776

[9] Camara C, Peris-Lopez P, Safkhani M, Bagheri N. ECGsound for human identification. Biomed Signal Processing Control. 2022;**72**:103335

[10] Ibtehaz N et al. "EDITH: ECG biometrics aided by deep learning for reliable individual authentication," IEEE Trans. Emerg. Top. Comput. Intell. 2021. pp. 1-27

[11] Hammad M, Pławiak P, Wang K, Acharya UR. ResNet-Attention model for human authentication using ECG signals. Expert Systems. 2021;**38**(6):1-17

[12] Srivastva R, Singh YN. ECG analysis for human recognition using non-fiducial methods. IET Biometrics. 2019;**8**(5):295-305

[13] Rathore H, Al-Ali A, Mohamed A, Du X, Guizani M. DTW based Authentication for Wireless Medical Device Security. In: 2018 14th Int. Wirel. Commun. Mob. Comput. Conf. IWCMC. 2018. pp. 476-481

[14] Rehman A, Saqib NA, Danial SM, Ahmed SH. ECG based authentication for remote patient monitoring in IoT by wavelets and template matching. In: Proc. IEEE Int. Conf. Softw. Eng. Serv. Sci. ICSESS. 2018. pp. 91-94

[15] Ba-Hammam A, Alhulwah S, Altamimi M, Alshebeili S. Authentication using ECG signals. In: 2017 Int. Conf. Electr. Comput. Technol. Appl. ICECTA 2017. 2018. pp. 1-4

[16] Sung D, Kim J, Koh M, Park K. ECG authentication in post-exercise situation. In: Proc. Annu. Int. Conf. IEEE Eng. Med. Biol. Soc. EMBS. 2017. pp. 446-449

[17] Chamatidis I, Katsika A, Spathoulas G. Using deep learning neural networks for ECG based authentication. In: Proc. - Int. Carnahan Conf. Secur. Technol. 2017. pp. 1-6

[18] Parastesh Karegar F, Fallah A, Rashidi S. ECG based human authentication with using Generalized Hurst Exponent. In: 2017 25th Iran. Conf. Electr. Eng. 2017. pp. 34-38

[19] Salloum R, Kuo CCCJ. ECG-based biometrics using recurrent neural networks. In: 2017 ICASSP, IEEE International Conference on Acoustics, Speech and Signal Processing – Proceedings. 2017. pp. 2062-2066

[20] Kim DH, Park JS, Kim IY, Kim SI, Lee, JS. Personal recognition using geometric features in the phase space of electrocardiogram. In: 2017 IEEE Life Sci. Conf. LSC. 2017. pp. 198-201

[21] Rezgui D, Lachiri Z. Integrating EMD attributes for person identification from electrocardiographic signals. In: 2016 Conf. Adv. Technol. Signal Image Process, ATSIP. 2016. pp. 478-482

[22] Arteaga-Falconi JS, Al Osman H, El Saddik A. ECG Authentication for Mobile Devices. IEEE Transactions on Instrumentation and Measurement. 2016;**65**(3):591-600

[23] Huang P, Li B, Guo L, Jin Z, Chen Y. A robust and reusable ECG-based authentication and data encryption scheme for eHealth systems. In: 2016 IEEE Glob. Commun. Conf. GLOBECOM. 2016. pp. 1-6

[24] Sivaranjani DNRB. Securing patient's confidential information using ECG Steganography. In: 2017 2nd International Conference on Communication and Electronics Systems (ICCES). pp. 540-544

[25] Yin S, Bae C, Kim SJ, Seo JS. Designing ECG-based physical unclonable function for security of wearable devices. In: Proc. Annu. Int. Conf. IEEE Eng. Med. Biol. Soc. EMBS. 2017. pp. 3509-3512

[26] Iancu-Constantin R, Serbanati LD, Chera C, Gheorghe-Pop ID, Ertl B. An E-health approach for remote cardiac rehabilitation. In: Proc. - 2015 20th Int. Conf. Control Syst. Comput. Sci. CSCS. 2015. pp. 205-210

[27] Khalifa W, Salem A, Roushdy M. A survey of EEG based user authentication schemes. In: 8th Int. Conf. INFOrmatics Syst, 14-16 May Bio-inspired Optim. Algorithms Their Appl. Track. 2012. pp. 55-60

[28] Abbas SN, Abo-Zahhad M, Ahmed SM. State-of-the-art methods and future perspectives for personal recognition based on electroencephalogram signals. IET Biometrics. 2015;**4**(3):179-190

[29] Revett K, Deravi F, Sirlantzis K. Biosignals for user authentication - Towards cognitive biometrics. In: 2010 Int. Conf. Emerg. Secur. Technol. ROBOSEC 2010 - Robot. Secur. LAB-RS 2010 - Learn. Adapt. Behav. Robot. Syst. 2010. pp. 71-76

[30] Sooriyaarachchi J, Seneviratne S, Thilakarathna K, Zomaya AY. MusicID: A brainwave-based user authentication system for internet of things. IEEE Internet of Things Journal. 2021;**8**(10):8304-8313

[31] Chen Y, Yin J. Design of electroencephalogram authentication access control to smart car. Healthcare Technology Letters. 2020;**7**(4):109-113

[32] Cauthen JM, Gandre T, Espinoza MAM, Patel MJ, Husain MI. An

authentication system using neurological responses to music. In: Proceedings - 2019 IEEE International Conference on Big Data. 2019. pp. 6001-6003

[33] Abdi Z, Alyasseri A, Khader AT, Al-betar MA, Alomari OA. EEG-based person authentication using multi-objective flower pollination algorithm. In: IEEE Congress on Evolutionary Computation (CEC). 2018

[34] Ong ZY, Ibrahim Z. Power spectral density analysis for human EEG- based biometric identification. In: Int. Conf. Comput. Approach Smart Syst. Des. Appl. 2018. pp. 1-6

[35] Khalafallah A, Ibrahim A, Shehab B, Raslan H, Eltobgy O, Elbaroudy S. A pragmatic authentication system using electroencephalography signals. In: 2018 IEEE International Conference on Acoustics, Speech and Signal Processing (ICASSP). 2018. pp. 901-905

[36] Arnau-Gonzalez P, Arevalillo-Herraez M, Katsigiannis S, Ramzan N. On the influence of affect in EEG-based subject identification. IEEE Transactions on Affective Computing. 2018;**3045**:1-11

[37] Nakamura T, Goverdovsky V, Mandic DP. In-ear EEG biometrics for feasible and readily collectable real-world person authentication. IEEE Transactions on Information Forensics and Security. 2018;**13**(3):648-661

[38] Wu Q, Yan B, Zeng Y, Zhang C, Tong L. Anti-deception: Reliable EEG-based biometrics with real-time capability from the neural response of face rapid serial visual presentation. Biomedical Engineering Online. 2018;**17**(1):1-16

[39] Kaur B, Singh D, Roy PP. A Novel framework of EEG-based user identification by analyzing music-listening behavior. Multimedia Tools and Applications. 2017;**76**(24):25581-25602

[40] Thomas KP, Vinod AP. EEG-Based Biometric Authentication Using Gamma Band Power During Rest State. Circuits, Systems, and Signal Processing. 2018;**37**(1):277-289

[41] Hasan M, Sohag HA, Ali E, Ahmad M. Estimation of the most effective rhythm for human identification using EEG signal. In: Proc. 9th Int. Conf. Electr. Comput. Eng. ICECE 2016. 2017. pp. 90-93

[42] Vahid A, Arbabi E. Human identification with EEG signals in different emotional states. In: 2016 23rd Iran. Conf. Biomed. Eng. 2016 1st Int. Iran. Conf. Biomed. Eng. ICBME 2016. 2017. pp. 242-246

[43] Dehzangi O, Williams C. Towards multi-modal wearable driver monitoring: Impact of road condition on driver distraction. In: 2015 IEEE 12th Int. Conf. Wearable Implant. Body Sens. Networks. 2015. pp. 1-6

[44] Abo-Zahhad M, Ahmed SM, Abbas SN. A Novel Biometric Approach for Human Identification and Verification Using Eye Blinking Signal. IEEE Signal Processing Letters. Jul. 2015;**22**(7):876-880

[45] Attarian U, Jain JU, Sadi SU, Mistry PU. Real time authentication based on blood flow parameters. 2018.

Chapter 11

A Voice Signal Filtering Methods for Speaker Biometric Identification

Eugene Fedorov, Tetyana Utkina and Tetyana Neskorodeva

Abstract

The preliminary stage of the personality biometric identification on a voice is voice signal filtering. For biometric identification are considered and in number investigated the following methods of noise suppression in a voice signal. The smoothing adaptive linear time filtering (algorithm of the minimum root mean square error, an algorithm of recursive least squares, an algorithm of Kalman filtering, a Lee algorithm), the smoothing adaptive linear frequency filtering (the generalized method, the MLEE (maximum likelihood envelope estimation) method, a wavelet analysis with threshold processing (universal threshold, SURE (Stein's Unbiased Risk Estimator)-threshold, minimax threshold, FDR (False Discovery Rate)-threshold, Bayesian threshold were used), the smoothing non-adaptive linear time filtering (the arithmetic mean filter, the normalized Gauss's filter, the normalized binomial filter), the smoothing nonlinear filtering (geometric mean filter, the harmonic mean filter, the contraharmonic filter, the α-trimmed mean filter, the median filter, the rank filter, the midpoint filter, the conservative filter, the morphological filter). Results of a numerical research of denoising methods for voice signals people from the TIMIT (Texas Instruments and Massachusetts Institute of Technology) database which were noise an additive Gaussian noise and multiplicative Gaussian noise were received.

Keywords: announcer biometric identification, voice signal filtering methods, the smoothing adaptive linear filtering, a wavelet analysis threshold processing, the smoothing non-adaptive linear filtering, the smoothing nonlinear filtering

1. Introduction

The preliminary stage of the personality biometric identification on a voice is voice signal filtering. Methods of a signal cleaning from noise arose and gained broad development in the twentieth century. With development a wavelet analysis joined normal time and frequency filters a wavelet filter.

Noise (interference) is the sound of an undesirable additional source added to a desired signal during its record or transfer on communication channel.

Noise can be classified by the following features: periodicity/aperiodicity; additive/multiplicative; continuity/impulsivity; to band width in a signal spectrum; color.

By continuity/impulsivity, noises are divided into: continuous; pulse (point); continuous and pulse.

Noises are divided by band width on:

- narrowband (noise with a continuous spectrum less than one octave);
- broadband (noise with a continuous spectrum more than one octave).

From color noise by the most difficult for filtering the white noise which has a uniform energy spectrum in all frequency range. The most widespread kind of a white noise is Gauss's noise.

Additive and multiplicative continuous and continuous impulse noises are removed from a signal by means of a wavelet analysis with threshold processing, the smoothing linear and many nonlinear filters, spectral subtraction. Impulse noises are removed many smoothing nonlinear filters. Additive aperiodic noise is removed low-frequency filters. Additive periodic noise is removed the bandpass and rejection filters.

2. The smoothing adaptive linear time filtering

Adaptive linear time filters call linear filters with adaptive impulse response function [1–3].

2.1 Algorithm of the minimum root mean square error

Algorithm of the minimum root mean square error which is applied to a signal $x(n)$ size N, is as follows [1]:

1. Impulse response function initialization

$$\mathbf{h} = \begin{pmatrix} h_1 \\ \dots \\ h_{2M+1} \end{pmatrix} = \begin{pmatrix} h(-M) \\ \dots \\ h(M) \end{pmatrix} = \begin{pmatrix} 0 \\ \dots \\ 0 \end{pmatrix}. \tag{1}$$

2. $n = M$.

3. Noise vector forming from a noise signal

$$\mathbf{v} = \begin{pmatrix} v_1 \\ \dots \\ v_{2M+1} \end{pmatrix} = \begin{pmatrix} v(n-M) \\ \dots \\ v(n+M) \end{pmatrix}. \tag{2}$$

4. Signal filtering (receiving noise estimates)

$$\breve{z}(n) = \mathbf{h}^T \mathbf{v}. \tag{3}$$

5. Error signal current value calculation

$$e(n) = x(n) - \breve{z}(n). \tag{4}$$

6. Impulse response function adaptation

$$\mathbf{h} = \mathbf{h} + \mu \mathbf{v} e(n), \tag{5}$$

where $0 < \mu < 1$.

7. If $n < N - M$ then $n = n + 1$, go to a step 2.

The signal is algorithm work result $e(n)$, $e(n) \approx s(n)$.

2.2 Recursive least squares algorithm

Recursive least squares algorithm which is applied to a signal $x(n)$ size N, is as follows [1]:

1. Initialization of impulse response function and adaptation matrix

$$\mathbf{h} = \begin{pmatrix} h_1 \\ \dots \\ h_{2M+1} \end{pmatrix} = \begin{pmatrix} h(-M) \\ \dots \\ h(M) \end{pmatrix} = \begin{pmatrix} 0 \\ \dots \\ 0 \end{pmatrix}, \mathbf{P} = \begin{pmatrix} p_{11} & \dots & p_{1,2M+1} \\ \dots & \dots & \dots \\ p_{2M+1,1} & \dots & p_{2M+1,2M+1} \end{pmatrix} = \lambda \mathbf{I}, \tag{6}$$

where λ-regularization parameter which is small at a big ratio signal/noise and is big at a small ratio signal/noise.

1. $n = M$.

2. Noise vector forming

$$\mathbf{v} = \begin{pmatrix} v_1 \\ \dots \\ v_{2M+1} \end{pmatrix} = \begin{pmatrix} v(n-M) \\ \dots \\ v(n+M) \end{pmatrix}. \tag{7}$$

3. Signal filtering (receiving noise estimates)

$$\breve{z}(n) = \mathbf{h}^T \mathbf{v}. \tag{8}$$

4. Error signal current value calculation

$$e(n) = x(n) - \breve{z}(n). \tag{9}$$

5. Adaptive gain $\mathbf{\Gamma}$ vector calculation

$$\mathbf{\Gamma} = \frac{\mathbf{P}\mathbf{v}}{\mathbf{v}^T \mathbf{P} \mathbf{v} + r}, \tag{10}$$

where $0 < r < 1$.

6. Estimates covariance matrix **P** calculation

$$\mathbf{P} = \frac{1}{r}\left(\mathbf{P} - \frac{\mathbf{P}\mathbf{v}\mathbf{v}^T\mathbf{P}}{\mathbf{v}^T\mathbf{P}\mathbf{v} + r}\right). \tag{11}$$

7. Impulse response function calculation

$$\mathbf{h} = \mathbf{h} + \mathbf{\Gamma} e(n). \tag{12}$$

8. If $n < N - M$ then $n = n + 1$, go to a step 2.

The signal is algorithm work result $e(n)$, $e(n) \approx s(n)$.

2.3 Kalman filtering algorithm

Kalman filtering algorithm which is applied to a signal $x(n)$ size N, is as follows [1]:

1. Estimates and white noise covariance matrixes:

$$\mathbf{h} = \begin{pmatrix} h_1 \\ \dots \\ h_{2M+1} \end{pmatrix} = \begin{pmatrix} h(-M) \\ \dots \\ h(M) \end{pmatrix} = \begin{pmatrix} 0 \\ \dots \\ 0 \end{pmatrix}, \mathbf{P} = \begin{pmatrix} p_{11} & \dots & p_{1,2M+1} \\ \dots & \dots & \dots \\ p_{2M+1,1} & \dots & p_{2M+1,2M+1} \end{pmatrix} = \lambda \mathbf{I}, \tag{13}$$

$$\mathbf{Q} = \begin{pmatrix} q_{11} & \dots & q_{1,2M+1} \\ \dots & \dots & \dots \\ q_{2M+1,1} & \dots & q_{2M+1,2M+1} \end{pmatrix} = \sigma_1^2 \mathbf{I},$$

where λ—regularization parameter which is small at a big ratio signal/noise and is big at a small ratio signal/noise,
σ_1^2—variance of a white noise of process which has null mean value.

1. $n = M$.

2. Noise vector forming from a noise signal

$$\mathbf{v} = \begin{pmatrix} v_1 \\ \dots \\ v_{2M+1} \end{pmatrix} = \begin{pmatrix} v(n-M) \\ \dots \\ v(n+M) \end{pmatrix}. \tag{14}$$

3. Signal filtering (receiving noise estimates)

$$\breve{z}(n) = \mathbf{h}^T \mathbf{v}. \tag{15}$$

4. Error signal current value calculation

$$e(n) = x(n) - \breve{z}(n). \tag{16}$$

5. Adaptive gain $\mathbf{\Gamma}$ vector calculation

$$\mathbf{\Gamma} = \frac{\mathbf{Pv}}{\mathbf{v}^T\mathbf{Pv} + \sigma_2^2}, \tag{17}$$

where σ_2^2—variance of a white noise of measurement which has null mean value.

6. Estimates covariance matrix $\mathbf{P}$ calculation

$$\mathbf{P} = \mathbf{P} - \frac{\mathbf{Pvv}^T\mathbf{P}}{\mathbf{v}^T\mathbf{Pv} + \sigma_2^2} + \mathbf{Q}. \tag{18}$$

7. Impulse response function calculation

$$\mathbf{h} = \mathbf{h} + \mathbf{\Gamma} e(n). \tag{19}$$

8. If $n < N - M$ then $n = n + 1$, go to a step 2.

The signal is algorithm work result $e(n)$, $e(n) \approx s(n)$.

2.4 Lee algorithm

Lee algorithm [2] which is applied to a signal $x(n)$ size N, is as follows:

1. Calculate local mean for each signal sample

$$\mu(n) = \frac{1}{2M+1}\sum_{m \in U_n} x(m), n \in \overline{M, N-M+1}, \tag{20}$$

where U_n—sample n neighborhood size $2M + 1$.

2. Calculate local variance for each signal sample

$$\sigma_x^2(n) = \frac{1}{2M+1}\sum_{m \in U_n} x^2(m) - \mu^2(m), n \in \overline{M, N-M+1}. \tag{21}$$

3. Calculate variance for each signal sample

$$\sigma_\nu^2 = \frac{1}{N-2M}\sum_{n} \sigma_x^2(n), n \in \overline{M, N-M+1}. \tag{22}$$

4. Execute adaptive filtering of a signal

$$s(n) = \sum_{m=-M}^{M} h(m)\ x(n-m), n \in \overline{M, N-M+1}. \tag{23}$$

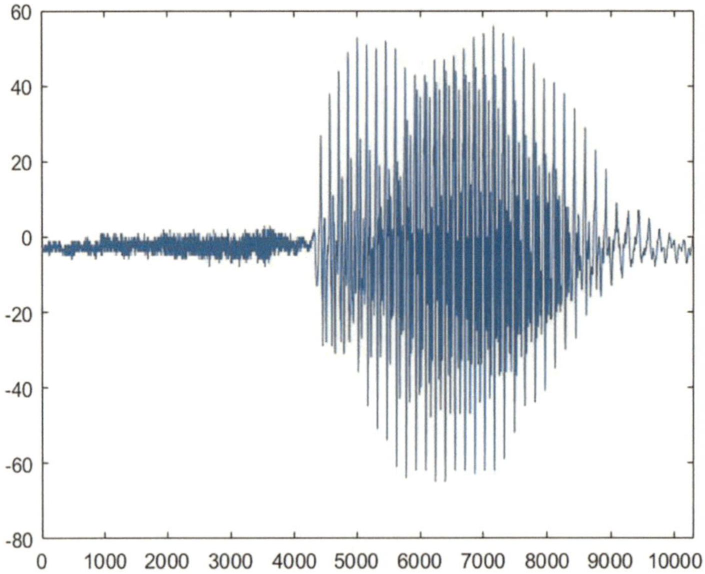

Figure 1.
Source signal for smoothing adaptive linear time filtering.

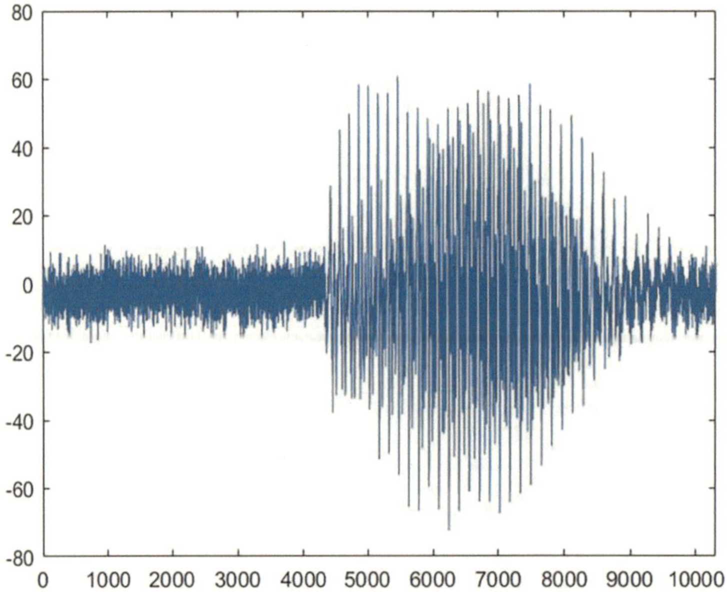

Figure 2.
A signal with an additive Gaussian noise for smoothing adaptive linear time filtering.

$$h(m) = \begin{cases} \dfrac{1}{2M+1} + \dfrac{\max\left\{0, \sigma_x^2(n) - \sigma_\nu^2\right\}}{\sigma_x^2(n)}\left(1 - \dfrac{1}{2M+1}\right), & m = 0 \\ \dfrac{1}{2M+1} - \dfrac{\max\left\{0, \sigma_x^2(n) - \sigma_\nu^2\right\}}{\sigma_x^2(n)} \cdot \dfrac{1}{2M+1}, & \textit{otherwise} \end{cases}. \quad (24)$$

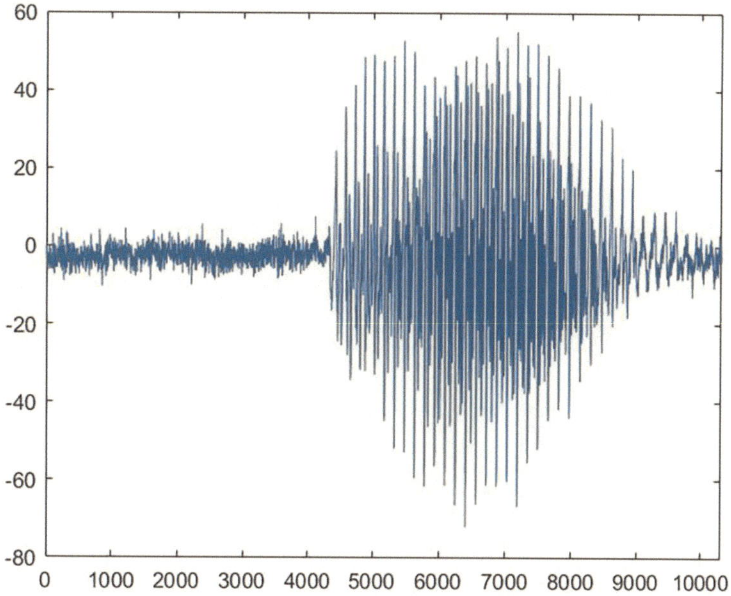

Figure 3.
The signal denoised by the adaptive filter.

Example

In **Figure 1** the source signal, is presented on **Figure 2**—noisy (additive white is added the noise with a mean 0 and variance 0.001 is Gaussian), on **Figure 3**—filtered and $M = 1$.

3. The smoothing adaptive linear frequency filtering

Adaptive linear frequency filters call linear filters with adaptive transfer function [4].

The smoothing adaptive linear frequency filtering is called *spectral subtraction.*

Let $X_p(k)$—a noisy signal spectrum of on p-th a frame, $V(k)$—mean noise spectrum, $S_p(k)$—a mean of the restored signal on p-th a frame.

Adaptive linear frequency filtering represents the inverse discrete Fourier transform of performing adaptive transfer function of the filter $H_p(k)$ on p-th frame and signal spectrum $X_p(k)$ on p-th a frame in a next form

$$y(n) = \frac{1}{N}\sum_{k=0}^{N-1}(X(k)H(k))e^{j\frac{2\pi nk}{N}}. \tag{25}$$

The following spectral subtraction methods are selected [1]:

1. General method (proposed Beruti, Schwartz and Makhoul)

$$S_p(k) = H_p(k)X_p(k), \tag{26}$$

$$H_p(k) = \begin{cases} G\left(\frac{|X_p(k)|^{\gamma} - \alpha|V(k)|^{\gamma}}{|X_p(k)|^{\gamma}}\right)^{1/\gamma}, & G\left(\frac{|X_p(k)|^{\gamma} - \alpha|V(k)|^{\gamma}}{|X_p(k)|^{\gamma}}\right)^{1/\gamma\gamma} > \beta|V(k)|, \\ \beta|V(k)|, & \textit{otherwise} \end{cases} \quad (27)$$

where $G, \ \alpha, \beta, \gamma$—parameters.

2. The Ball method

$$H_p(k) = \begin{cases} \frac{|X_p(k)| - |V(k)|}{|X_p(k)|}, & |X_p(k)| - |V(k)| > 0 \\ 0, & \textit{otherwise} \end{cases}. \quad (28)$$

3. Wiener filtering

$$H_p(k) = \begin{cases} \frac{|X_p(k)|^2 - |V(k)|^2}{|X_p(k)|^2}, & |X_p(k)|^2 - |V(k)|^2 > 0 \\ 0, & \textit{otherwise} \end{cases}. \quad (29)$$

4. The MLEE method

$$S_p(k) = H_p(k)X_p(k),$$

$$H_p(k) = \begin{cases} \frac{1}{2} + \frac{1}{2}\sqrt{\frac{|X_p(k)|^2 - |V(k)|^2}{|X_p(k)|^2}}, & |X_p(k)|^2 - |V(k)|^2 > 0 \\ 0, & \textit{otherwise} \end{cases}. \quad (30)$$

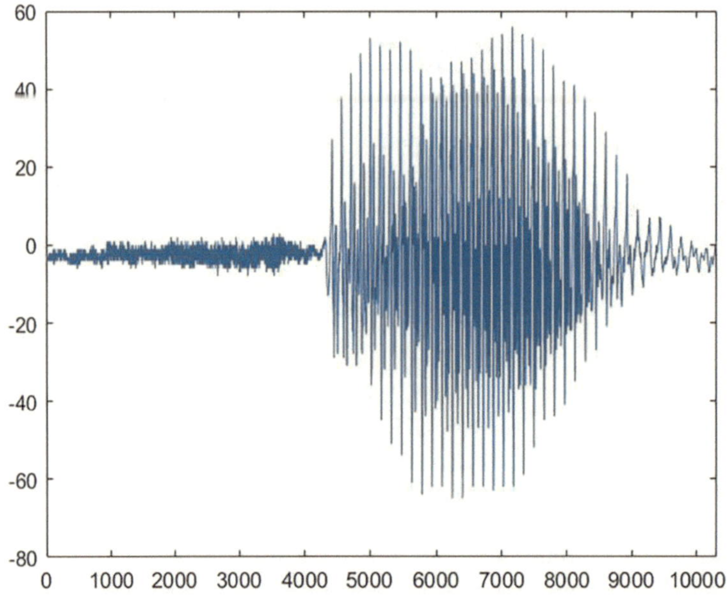

Figure 4.
Source signal for smoothing adaptive linear frequency filtering.

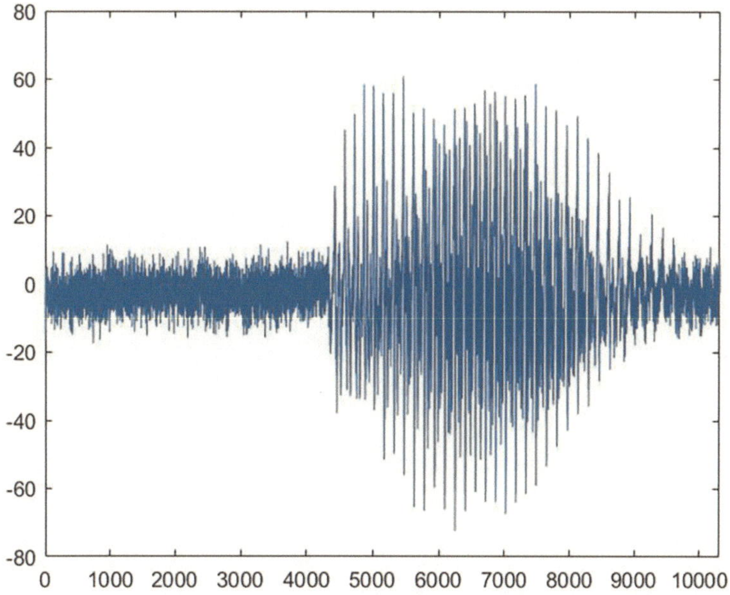

Figure 5.
A signal with additive Gaussian noise for smoothing adaptive linear frequency filtering.

Example

In **Figure 4** the source signal, is presented on **Figure 5**—noisy (additive white is added the noise with a mean 0 and variance 0.001 is Gaussian), the signals denoised by means of filtering according to general method ($G = 1$, $\gamma = 2$, $\alpha = 6$, $\beta = 0.1$) (**Figure 6**), Ball (**Figure 7**), Wiener (**Figure 8**), MLEE (**Figure 9**). For these methods

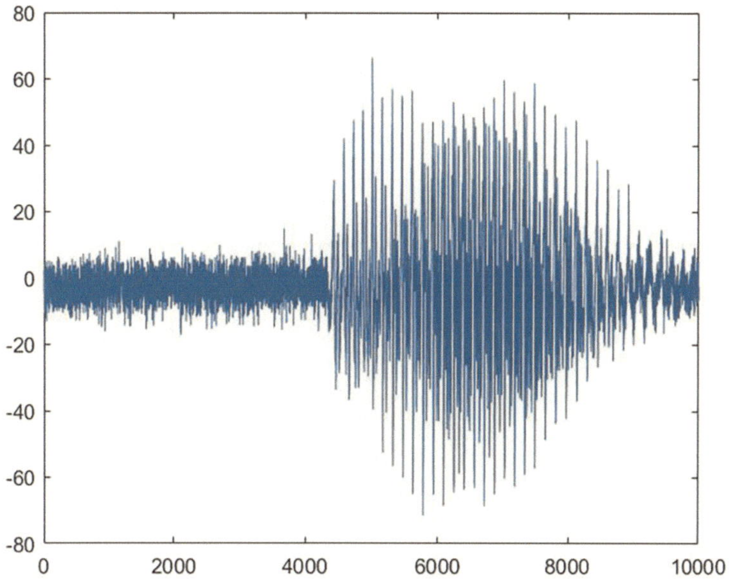

Figure 6.
The signal denoised by means of filtering according to general method.

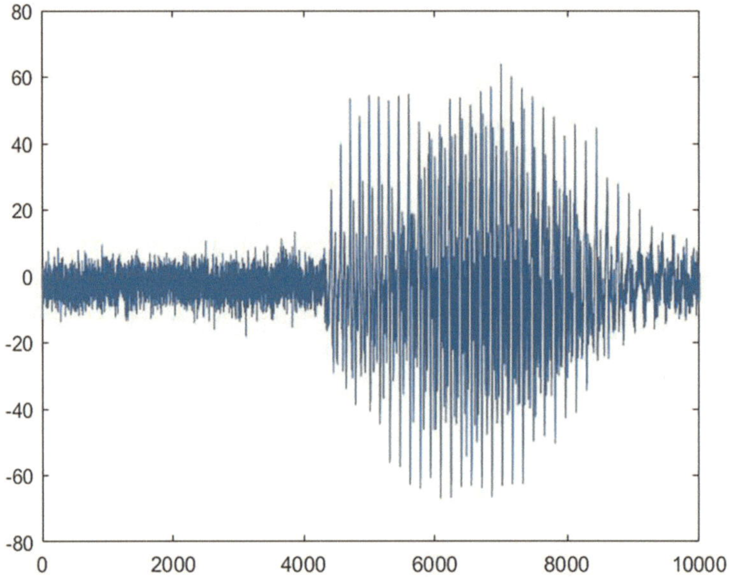

Figure 7.
The signal denoised by means of filtering according to Ball.

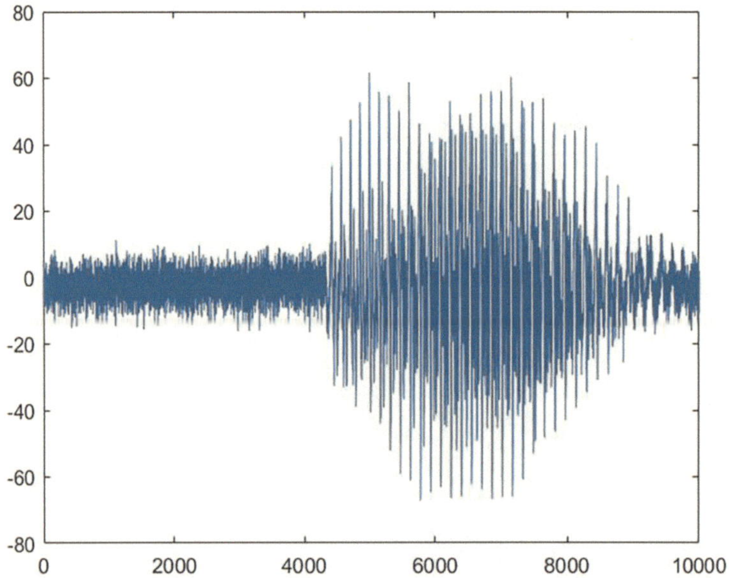

Figure 8.
The signal denoised by means of filtering according to Wiener.

as frame length, it was selected ΔN = 512 (about 20 ms). In signal quality the word "Sasha" with a sampling rate of 22050 Hz, 8-bits, mono was selected.

4. Wavelet analysis threshold processing

For a wavelet analysis the soft and rigid threshold processing is widely used [5].

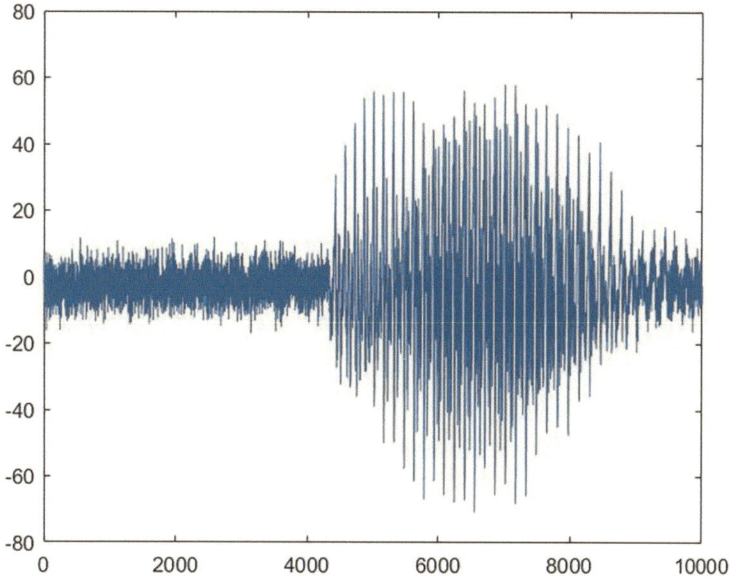

Figure 9.
The signal denoised by means of filtering according to MLEE.

4.1 Signal analysis

1. Initialization.

   Decompositions level number $i = 1$.

$$c_{0x} = s(x), x \in \overline{0, N/2^{i-1} - 1}, \quad (31)$$

   where $s(x)$—original signal length N.

2. On the current ith the decomposition level signal convolution with impulse response functions of FIR-HP (Finite Impulse Response—High Pass) and FIR-LP (Finite Impulse Response—Low Pass) filter is executed $g(k)$, $h(k)$ respectively

$$d_{im} = \sqrt{2} \sum_{k=0}^{N_2/2^{i-1}-1} c_{i-1,k} g(k + 2m), m \in \overline{0, N/2^i - 1}, \quad (32)$$

$$c_{im} = \sqrt{2} \sum_{k=0}^{N_2/2^{i-1}-1} c_{i-1,k} h(k + 2m), m \in \overline{0, N/2^i - 1}. \quad (33)$$

3. If $i < P$ then $i = i + 1$, go to a step 1.

4.2 Decomposition coefficients conversion

1. Decompositions level number $i = 1$.

2. Create the vector arranged on increase

$$a_i = \left(|d_{i0}|, \dots, \left|d_{i,N/2^i-1}\right|\right), |d_{im}| < |d_{i,m+1}|. \quad (34)$$

3. Calculate noise standard deviation based on a received vector median

$$\sigma_i = \frac{median(a_i)}{0.6745}, \quad (35)$$

where $median(x)$—function which returns a median of a vector x.

4. Calculate one of the following thresholds

1. Calculate a universal threshold

$$T_i = \sigma_i\sqrt{2\ln\left(N/2^i\right)}. \quad (36)$$

2. Calculate a SURE-threshold

1. Define a threshold based on minimal risk

$$r_{im} = 1 + \frac{-2(m-1) + \sum_{k=0}^{m-1}(a_{ik})^2 + (a_{im})^2\left(N/2^i - 1 - m\right)}{\left(N/2^i\right)}, m \in \overline{0, N/2^i - 1}, \quad (37)$$

$$m^* = \arg\min_m r_{im}, m \in \overline{0, N/2^i - 1}, \tilde{T}_i = a_{im^*}. \quad (38)$$

2. Calculate a SURE-threshold based on the received threshold

$$T_i = \begin{cases} \sigma_i\sqrt{2\ln\left(N/2^i\right)}, & \sum\limits_{m=0}^{N/2^i-1} d_{im} - \left(N/2^i\right)\sigma_i^2 \le \varepsilon_i \\ \tilde{T}_i, & \sum\limits_{m=0}^{N/2^i-1} d_{im} - \left(N/2^i\right)\sigma_i^2 > \varepsilon_i \end{cases}, \varepsilon_i = \sigma_i^2\sqrt{\left(N^f/2^i\right)\ln\left(N/2^i\right)^3}. \quad (39)$$

3. Calculate a minimax threshold

$$T_i = \begin{cases} \sigma_i\left(0.3936 + 0.1829\ln\left(N/2^i\right)\right), & N/2^i > 32 \\ 0, & N/2^i \le 32 \end{cases}. \quad (40)$$

4. Calculate a FDR-threshold

$$\mu_i = \frac{\sum_{m=1}^{N/2^i-1} a_{im}}{N/2^i - 1}, \Delta_{im} = erfc\left(\frac{1}{\sqrt{2}}\left|\frac{a_{im} - \mu_i}{\sigma_i}\right|\right) - q\frac{m}{N/2^i - 1}, m \in \overline{0, N/2^i - 1}, \quad (41)$$

$$m^* = \arg\min_{m} \left(sgn\left(\Delta_{im}\right) sgn\left(\Delta_{i,m+1}\right)\right), m \in \overline{0, N/2^i - 2}, T_i = a_{im^*}, \tag{42}$$

where q—parameter, $q \in (0, 0.5]$, and it is normal $q = 0.05$, $erfc(x)$—additional function of errors.

5. Calculate a Bayesian threshold (using Quasi-Cauchy distribution which is the most effective)

   1. Calculate function β (using Quasi-Cauchy distribution)

$$\beta(a_{im}) = \frac{g(a_{im})}{\varphi(a_{im})} - 1, m \in \overline{0, N/2^i - 1}, g(x) = (\gamma \cdot \varphi)(x) = \frac{1}{\sqrt{2\pi}} x^{-2}\left(1 - e^{-x^2/2}\right), \tag{43}$$

$$\varphi(x) = \frac{1}{\sqrt{2\pi}} e^{-x^2/2}, \gamma(x) = \frac{1}{\sqrt{2\pi}} \frac{\varphi(x) - |x|(1 - \Phi(x))}{\varphi(x)}, \tag{44}$$

where $\varphi(x)$—standard normal distribution density, $\gamma(x)$—Quasi-Cauchy's density of distribution.

   2. Calculate the minimum parameter w_i value (using Quasi-Cauchy distribution)

$$w_i^{\min} = \frac{\frac{1}{2}\left(\tilde{T}_i\right)^2 e^{-\left(\tilde{T}_i\right)^2/2}}{1 + \Phi\left(\tilde{T}_i\right) - \tilde{T}_i \ \varphi\left(\tilde{T}_i\right) - \frac{1}{2}}, \tilde{T}_i = \sqrt{2\ln\left(N/2^i\right)}. \tag{45}$$

   3. Find parameter value w_i by the equation numerical solution on an interval $\left[w_i^{\min}, 1\right]$

$$S_i(w_i) = \sum_{m=0}^{N/2^i - 1} \frac{\beta(a_{im})}{1 + w_i \beta(a_{im})} = 0. \tag{46}$$

   4. Find a Bayesian threshold T_i by the numerical solution of the equation on an interval $[0, T^{\max}]$ (using Quasi-Cauchy distribution)

$$-\Phi(T_i) + T_i \ \varphi(T_i) + \frac{1}{2} + \frac{1}{2}(T_i)^2 e^{-(T_i)^2/2}(1/w_i - 1) = 0. \tag{47}$$

5. Execute one of the following thresholds processing

   1. Execute soft threshold processing (for universal, minimax, Bayesian, a SURE-threshold)

$$\tilde{d}_{im} = \begin{cases} d_{im} - T_i, & d_{im} \ge T_i \\ d_{im} + T_i, & d_{im} \le -T_i \\ 0, & |d_{im}| \le T_i \end{cases}, m \in \overline{0, N/2^{i-1} - 1}. \tag{48}$$

2. Execute rigid threshold processing (for universal, minimax, Bayesian, SURE, a FDR threshold)

$$\tilde{d}_{im} = \begin{cases} d_{im}, & |d_{im}| > T_i \\ 0, & |d_{im}| \le T_i \end{cases}, m \in \overline{0, N/2^{i-1} - 1}. \quad (49)$$

6. If $i < P$ then $i = i + 1$, go to a step 1.

4.3 Signal design

1. Initialization.

   Level number of recoveries $i = P$.

2. On the current i-th the recovery level signal convolution with impulse response functions of FIR-HP and FIR-LP filter is executed $g(k)$, $h(k)$ respectively

$$c_{i-1,n} = \sqrt{2} \sum_{m=0}^{N/2^i - 1} c_{im} h(n + 2m) + \sqrt{2} \sum_{m=0}^{N/2^i - 1} \tilde{d}_{im} g(n + 2m), n \in \overline{0, N/2^{i-1} - 1}. \quad (50)$$

3. If $i > 1$ then $i = i - 1$, go to a step 1.

Example

In **Figure 10** the source signal, is presented on **Figure 11**—noisy (additive white is added the noise with a mean 0 and variance 0.001 is Gaussian), in **Figure 12**—filtered. The soft SURE-threshold with Daubechies wavelet with amount of the zero moments $L = 4$ was used (i.e., an order of FIR-HP and FIR-LP filter $M = 8$).

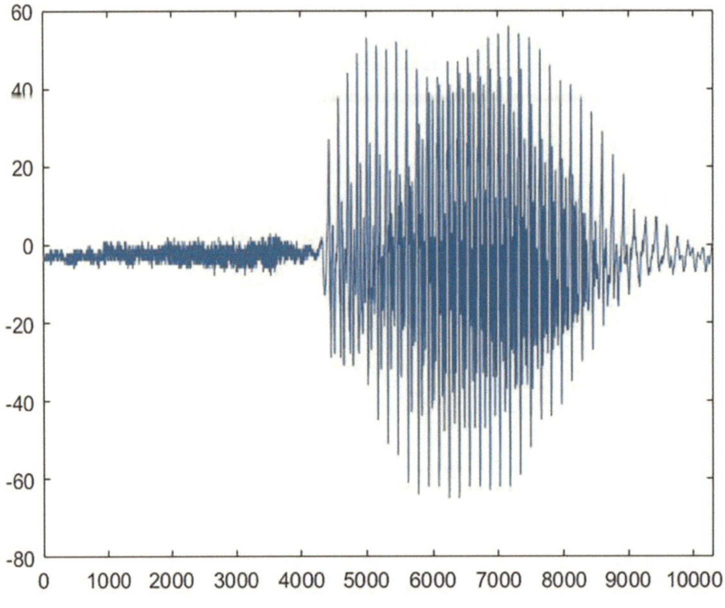

Figure 10.
Source signal for wavelet analysis threshold processing.

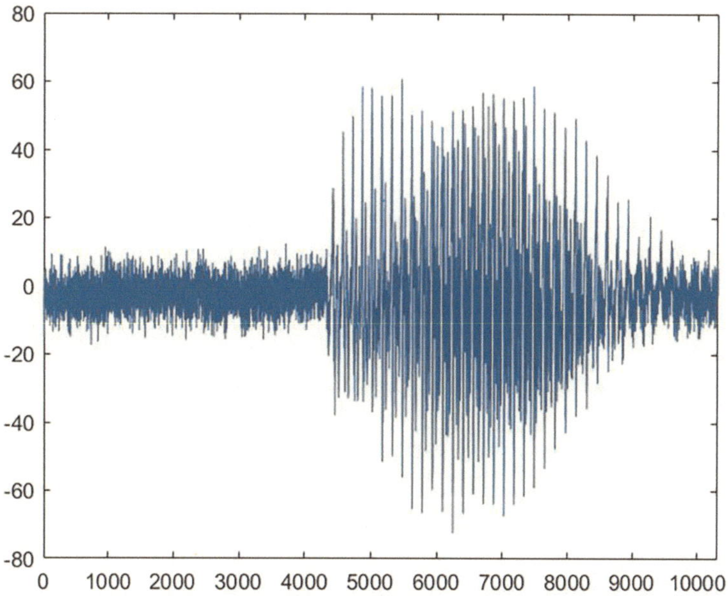

Figure 11.
A signal with an additive Gaussian noise for wavelet analysis threshold processing.

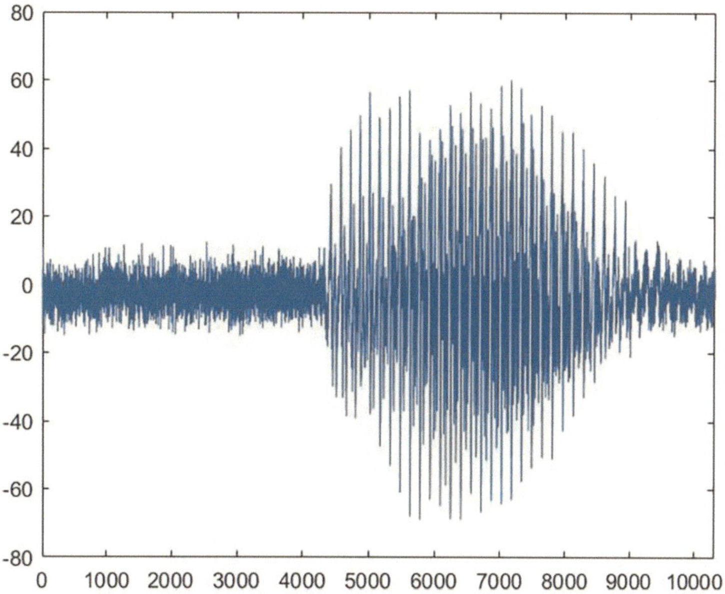

Figure 12.
The signal cleaned using a wavelet analysis with threshold processing.

5. The smoothing non-adaptive linear temporary filtering

The smoothing non-adaptive linear time filters are low pass filters [6].

In case of the FIR-LP filter with symmetric impulse response function $h(-M), \dots, h(0), \dots, h(M)$, non-adaptive linear time filtering represents convolution of non-adaptive impulse response function $h(m)$ signal $x(n)$ as

$$y(n) = \sum_{m=-M}^{M} h(m)\ x(n-m). \quad (51)$$

Let us give impulse response functions of the most widespread two-dimensional smoothing linear filters:

1. Impulse response function of the arithmetic mean filter

$$h(m) = \frac{1}{2M+1}, m \in \overline{-M, M}. \quad (52)$$

2. Impulse response function of the normalized Gauss filter

$$h(m) = \frac{\frac{1}{2\pi\sigma^2} \exp\left(-\frac{1}{2}\frac{m^2}{\sigma^2}\right)}{\sum_{l=-M}^{M} \frac{1}{2\pi\sigma^2} \exp\left(-\frac{1}{2}\frac{l^2}{\sigma^2}\right)}, m \in \overline{-M, M}. \quad (53)$$

3. Impulse response function of the normalized binomial filter

$$h(m) = \frac{C_{2M}^{M+m}}{\sum_{l=0}^{2M} C_{2M}^{l}}, C_n^m = \frac{n!}{m!(n-m)!}, m \in \overline{-M, M}. \quad (54)$$

Example

In **Figure 13** the source signal, is presented on **Figure 14**—noisy (additive white is added the noise with a mean 0 and variance 0.001 is Gaussian), on **Figure 15**—filtered, wherein the arithmetic mean filter with $M = 1$.

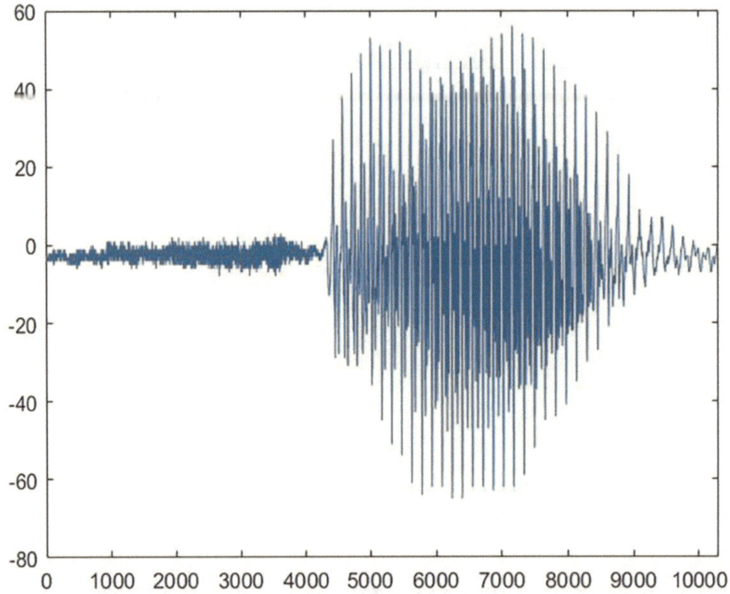

Figure 13.
Source signal for smoothing non-adaptive linear temporary filtering.

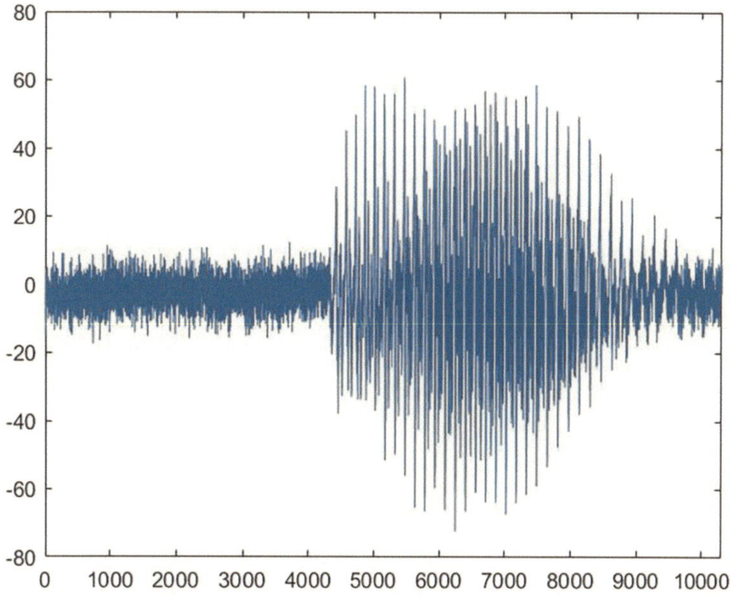

Figure 14.
A signal with an additive Gaussian noise for smoothing non-adaptive linear temporary filtering.

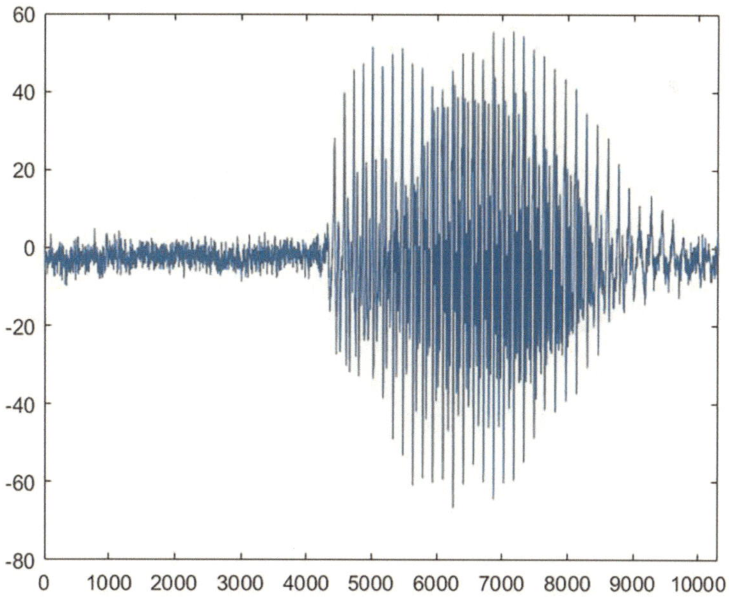

Figure 15.
The signal denoised by means of the arithmetic mean filter.

6. The smoothing nonlinear filtering

The smoothing nonlinear filters [6] are low-pass filters.

6.1 Geometric mean, harmonic mean, contraharmonic filters

1. Geometric mean filter

$$y(n) = \left(\prod_{m=-M}^{M} x(n-m) \right)^{\frac{1}{2M+1}}. \tag{55}$$

2. Harmonic mean filter

$$y(n) = \frac{2M+1}{\sum_{m=-M}^{M} \frac{1}{x(n-m)}}. \tag{56}$$

3. Contraharmonic filter

$$y(n) = \frac{\sum_{m=-M}^{M} x^{Q+1}(n-m)}{\sum_{m=-M}^{M} x^{Q}(n-m)}. \tag{57}$$

Geometric mean, harmonic mean, contraharmonic filters delete additive and multiplicative continuous and continuous impulse noises.

The harmonic mean filter in case of an impulse noise suppresses only white points.

The contraharmonic filter in case of an impulse noise at $Q > 0$ suppresses only black points (at $Q = -1$ receive the harmonic mean filter), and at $Q < 0$ suppresses only white points. At $Q = 0$ receive the arithmetic mean filter.

Therefore, for suppression of an impulse noise it is better to use α-trimmed mean, median or rank, conservative and morphological filters.

6.2 α-trimmed mean filter

Algorithm α-trimmed mean filtering applied to a signal $x(n)$ size N, is as follows:

1. Create for each sample of a signal a vector from elements of its neighborhood U_n size $2M + 1$ as

$$a_n = (x(n-M), \ldots, x(n+M)), n \in \overline{M, N-M+1}. \tag{58}$$

2. Sort for each sample of a signal element of its vector by increase

$$\tilde{a}_n = sort(a_n), n \in \overline{M, N-M+1}. \tag{59}$$

3. Execute α-trimmed mean filtering of a signal in a form

$$y(n) = \frac{\sum_{m=1+\alpha/2}^{2M+1+\alpha/2} \tilde{a}_n(m)}{2M+1-\alpha}, n \in \overline{M, N-M+1}, \tag{60}$$

where α—parameter, which multiple 2, $0 \leq \alpha \leq 2M$.

At $\alpha = 0$ we receive the arithmetic mean filter, and at $\alpha = 2M$ receive the median filter. α-trimmed mean filter deletes additive and multiplicative continuous both continuous impulse noises and impulse noises.

6.3 Median and rank filters

Median filtering is defined in a form.

$$y(n) = \underset{m \in U_n}{\text{median}}\{x(m)\}, n \in \overline{M, N - M + 1}. \quad (61)$$

where U_n—neighborhood of sample n size $2M + 1$.

Median filtering is a special case of rank filtering at a rank $r = M + 1$. In case of rank filtering not the central sample, but sample which number corresponds to a rank undertakes r, and $1 \le r \le 2M + 1$.

Median and rank filters delete additive and multiplicative continuous both continuous impulse noises and impulse noises.

6.4 Midpoint filter

Algorithm of the midpoint filtering applied to a signal $x(n)$ size N, is as follows:

1. Calculate for each sample of a signal the minimum and maximum value in its neighborhood in a form

$$\alpha(n) = \min_{m \in U_n}\{x(m)\}, \beta(n) = \max_{m \in U_n}\{x(m)\}, n \in \overline{M, N - M + 1}, \quad (62)$$

where U_n—neighborhood of sample n size $2M + 1$.

2. Signal midpoint filtering execute in a form

$$y(n) = \frac{1}{2}(\alpha(n) + \beta(n)), n \in \overline{M, N - M + 1}. \quad (63)$$

The Midpoint filter deletes additive and multiplicative continuous and continuous impulse noises.

6.5 Conservative filter algorithm

Conservative filtering algorithm applied to a signal $x(n)$ size N, is as follows:

1. Calculate for each sample of a signal the minimum and maximum value in its neighborhood in a form

$$\alpha(n) = \min_{m \in U_n \backslash \{n\}}\{x(m)\}, \beta(n) = \max_{m \in U_n \backslash \{n\}}\{x(m)\}, n \in \overline{M, N - M + 1}. \quad (64)$$

where U_n—sample neighborhood n size $2M + 1$.

2. Signal conservative filtering execute in a form

$$y(n)=\begin{cases} x(n), & \alpha(n)<x(n)<\beta(n) \\ \alpha(n), & x(n)\le\alpha(n) \\ \beta(n), & x(n)\ge\beta(n) \end{cases}, n\in\overline{M,N-M+1}. \quad (65)$$

The conservative filter deletes additive and multiplicative continuous both continuous impulse noises and impulse noises.

6.6 Morphological filter

Morphological filtering is carried out by consecutive performing operations of open and close or close and open. At open, operations dilatation and an erosion are consistently executed, and at close—an erosion and dilatation.

Dilatation can be defined in a form.

$$z(n)=\max_{m\in U_n}\{x(m)\}\ n\in\overline{M,N-M+1}. \quad (66)$$

Erosion can be defined in a form.

$$z(n)=\min_{m\in U_n}\{x(m)\}, n\in\overline{M,N-M+1}, \quad (67)$$

where U_n—sample neighborhood n.

The morphological filter deletes impulse noises.

Example

In **Figure 16** the source signal, is presented on **Figure 17**—noisy (additive white is added the noise with an mean 0 and variance 0.001 is Gaussian), the signals denoised

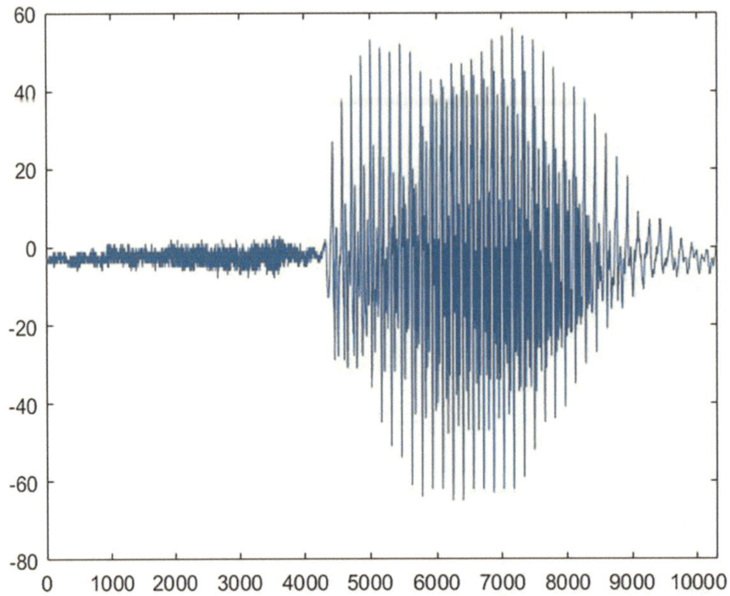

Figure 16.
Source signal for smoothing nonlinear filtering of additive Gaussian noise.

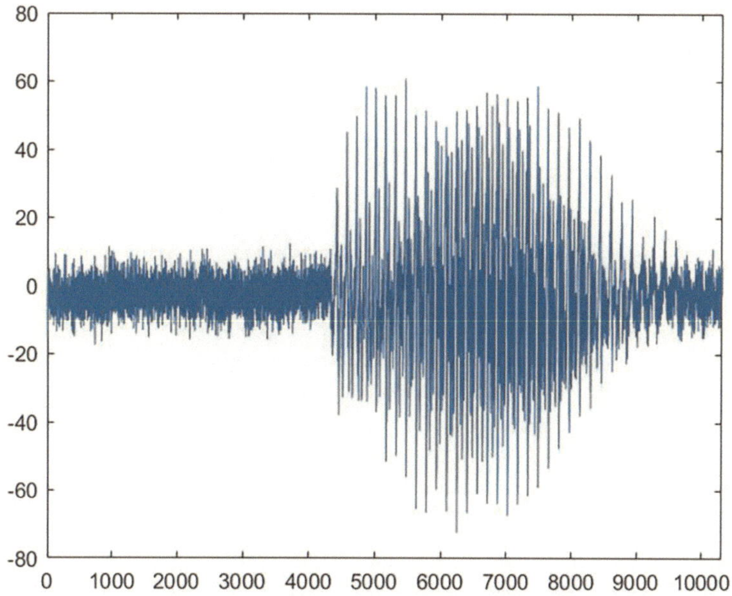

Figure 17.
A signal with an additive Gaussian noise for smoothing nonlinear filtering.

by means of the geometric mean filter ($M = 1$) (**Figure 18**), α-trimmed mean filter ($M = 2$, $\alpha = M = 2$) (**Figure 19**), median filter ($M = 2$) (**Figure 20**), midpoint filter ($M = 1$) (**Figure 21**), conservative filter ($M = 1$) (**Figure 22**). In signal quality the syllable "sa" with a sampling rate of 22050 Hz, 8-bits, mono was selected.

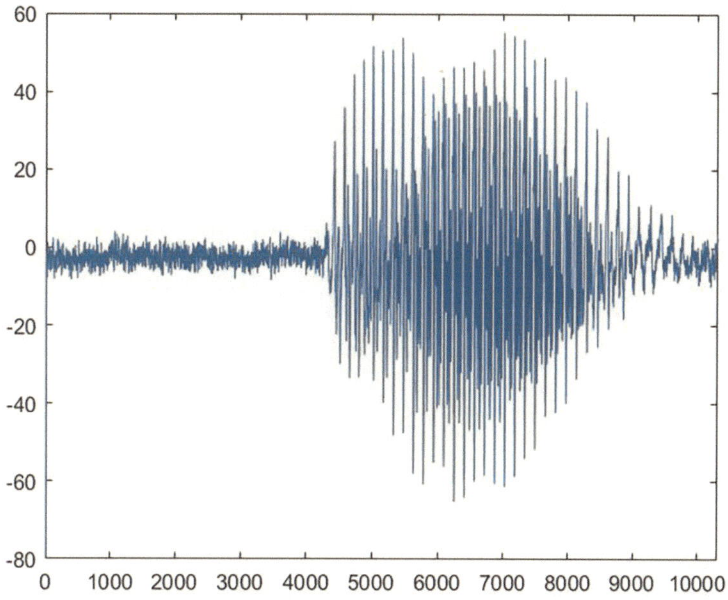

Figure 18.
The signal denoised by means of the geometric mean filter.

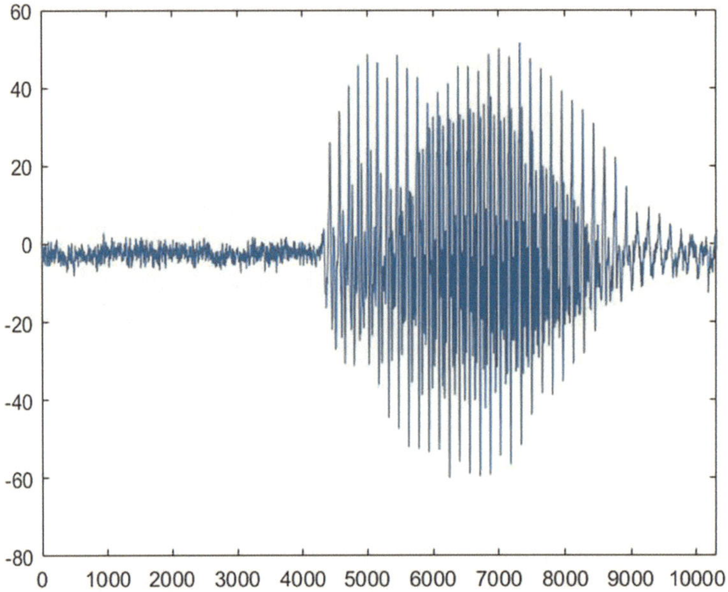

Figure 19.
The signal denoised by means of the α-trimmed mean filter.

Example

In **Figure 23** the source signal, on **Figure 24**—noisy (the impulse noise "salt and pepper" with a noisiness of 1% of sample of a signal is added), the signals denoised by means of the α-trimmed mean filter ($M = 2$, $\alpha = M = 2$) (**Figure 25**), the median

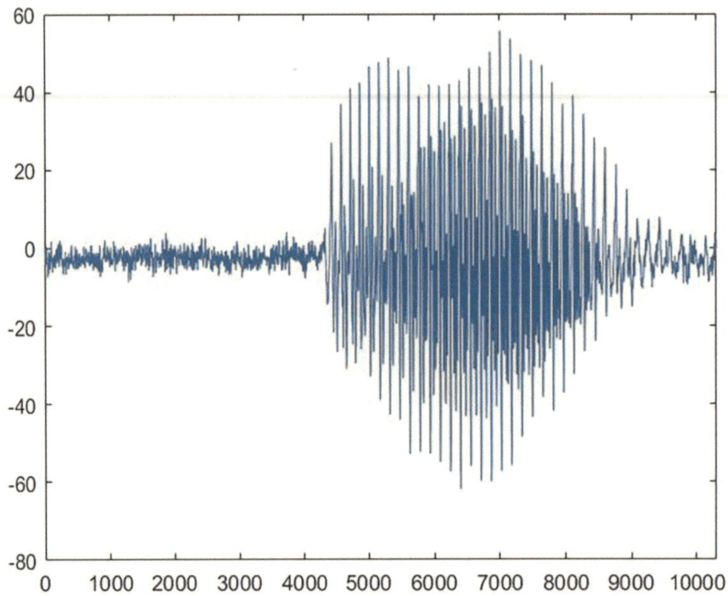

Figure 20.
The signal denoised by means of the median filter.

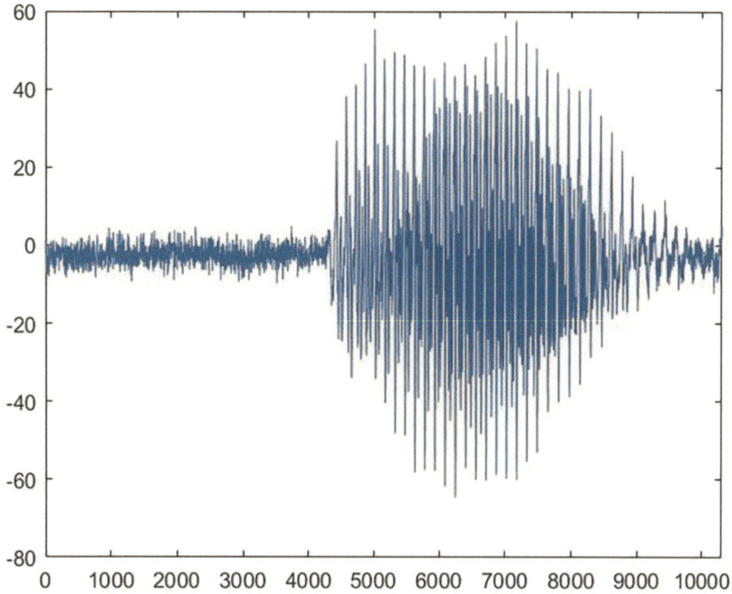

Figure 21.
The signal denoised by means of the midpoint filter.

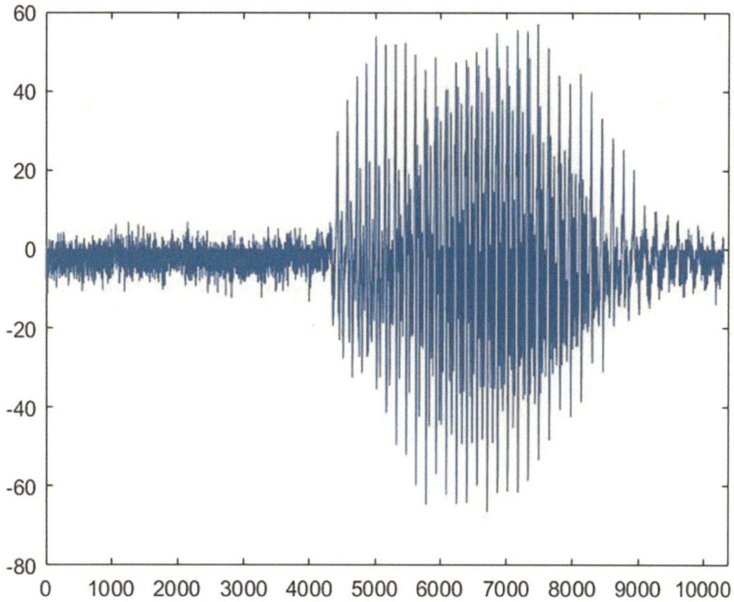

Figure 22.
The signal denoised by means of the conservative filter.

filter ($M = 2$) (**Figure 26**), the conservative filter ($M = 1$) (**Figure 27**), the morphological filter (consistently executed by open and close with $M = 3$) (**Figure 28**). In signal quality the syllable "sa" a sampling rate of 22050 Hz, 8-bits, mono was selected.

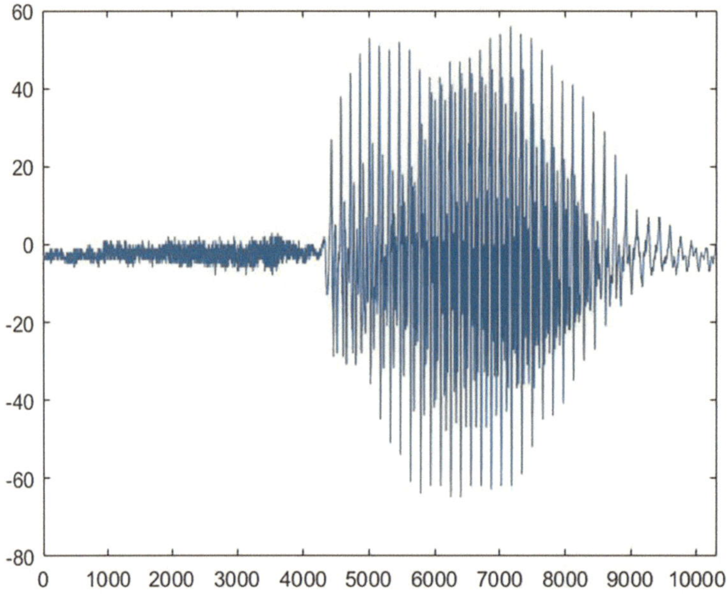

Figure 23.
Source signal for smoothing nonlinear filtering of impulse noise.

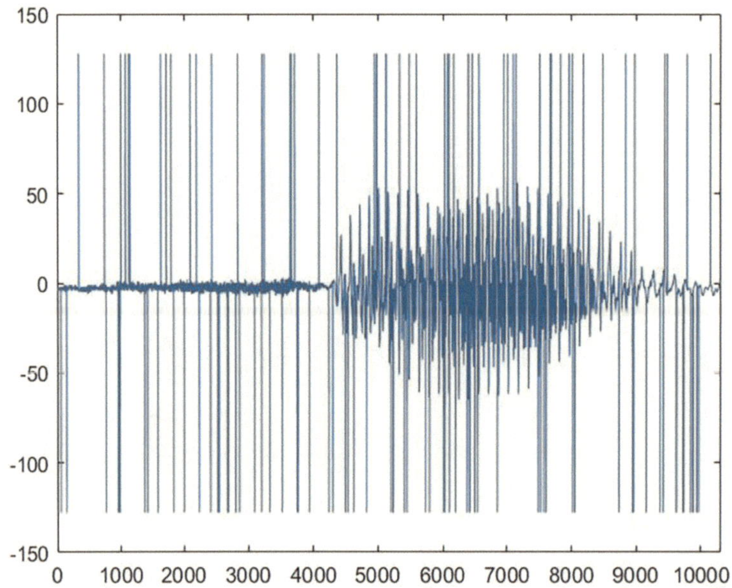

Figure 24.
A signal with an impulse noise "salt and pepper" for smoothing nonlinear filtering.

7. Numerical research of denoising methods noise

For the voice signals containing vocal sounds the sampling rate of 8 kHz and quantity of quantizing levels 256 was set.

Numerical research results of denoising methods on a basis a wavelet analysis with threshold processing in case of Daubechies wavelet about 8 with soft threshold

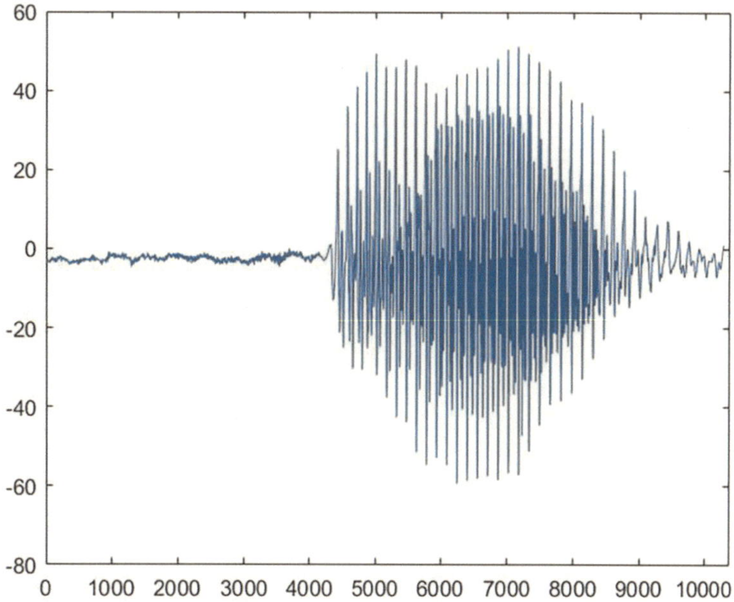

Figure 25.
The signal denoised by means of the α-trimmed mean filter.

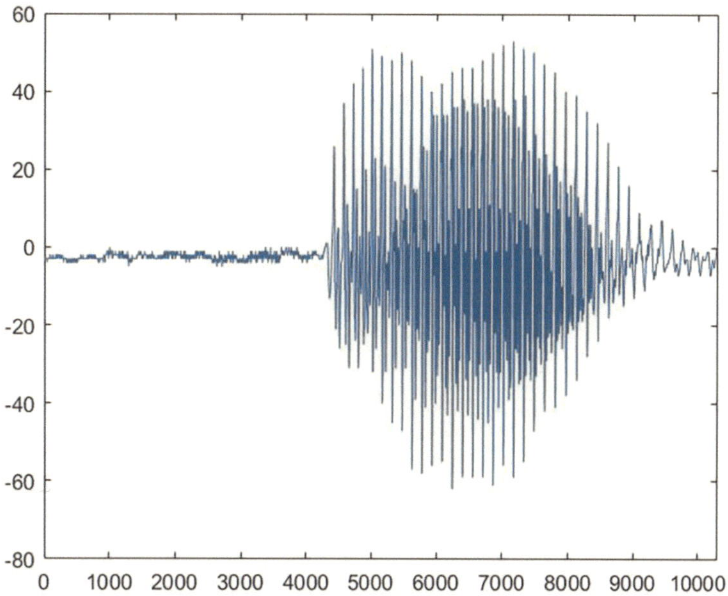

Figure 26.
The signal denoised by means of the median filter.

processing with a SURE-threshold, the adaptive filter about 1, it is Gaussian the filter about 1 with parameter $\sigma = 0.7$, the arithmetic mean filter about 1, geometric mean filters about 1, harmonic mean filters about 1, contraharmonic filters about 1 with parameter $Q = 1$, median filter about 2, α-trimmed mean filter of about 2 with parameter $\alpha = 2$, the midpoint filter about 1, conservative filters about 1 for voice

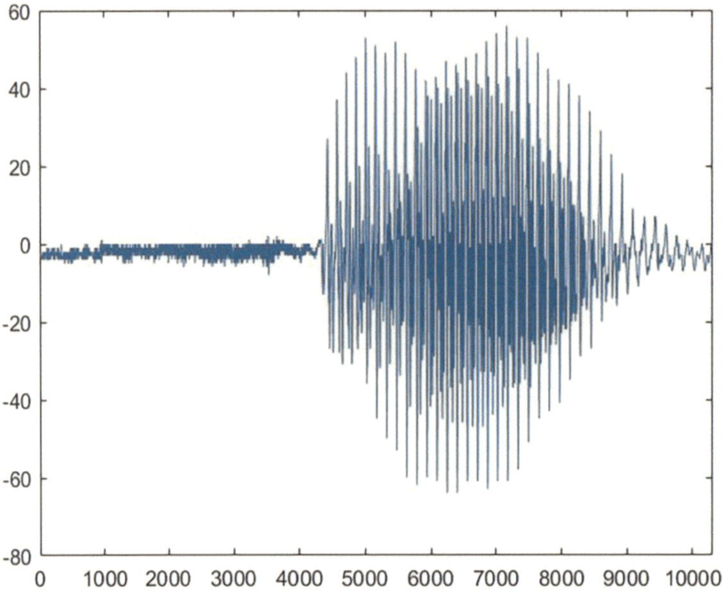

Figure 27.
The signal denoised by means of the conservative filter.

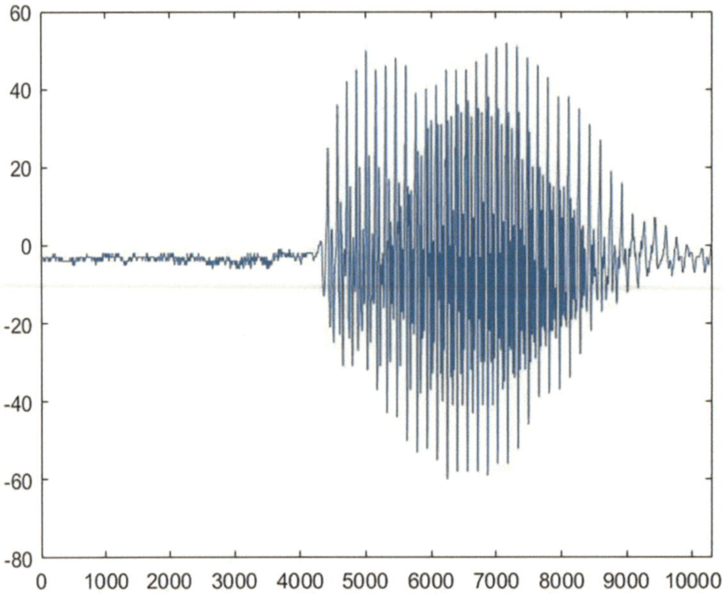

Figure 28.
The signal denoised by means of the morphological filter.

signals people from the TIMIT database which were noise an additive Gaussian noise with mean 0 and variance 0.001 (a signal-to-noise ratio about 11 dB) and multiplicative Gaussian noise with mean 1 and variance 0.07 (a signal-to-noise ratio about 23 dB), are presented to **Table 1**, where MSE—Mean Square Error.

Denoising method on a basis	MSE	
	Additive Gaussian noise	Multiplicative Gaussian noise
Wavelet analysis with threshold processing	11.2809	14.7860
Adaptive filter	9.9072	13.3914
Gaussian filter	14.3395	14.7662
Arithmetic mean filter	13.4738	14.0495
Geometric mean filter	15.5846	15.3252
Harmonic mean filter	20.4298	19.8623
Contraharmonic filter	13.3552	13.4845
Median filter	6.8697	6.5193
α-Trimmed mean filter	5.0843	4.9043
Midpoint filter	6.7667	6.4873
Conservative filter	9.0294	9.6437

Table 1.
Results of a numerical research of denoising methods from additive Gaussian noise and multiplicative Gaussian noise.

The result is provided in **Table 1** shows that the smallest MSE is provided α-trimmed mean filter.

8. Conclusion

For biometric identification are considered and in number investigated the following methods of noise suppression in a voice signal. The smoothing adaptive linear time filtering (the minimum root mean square error algorithm, the recursive least squares algorithm, the Kalman filtering algorithm, the Lee algorithm), the smoothing adaptive linear frequency filtering (the generalized method, the MLEE method, a wavelet analysis with threshold processing (universal threshold, SURE-threshold, minimax threshold, FDR-threshold, Bayesian threshold were used), the smoothing non-adaptive linear time filtering (the arithmetic mean filter, the normalized Gauss's filter, the normalized binomial filter), the smoothing nonlinear filtering (geometric mean filter, the harmonic mean filter, the contraharmonic filter, the α-trimmed mean filter, the median filter, the rank filter, the midpoint filter, the conservative filter, the morphological filter). Numerical research results of denoising methods for voice signals people from the TIMIT database which were noise an additive Gaussian noise and multiplicative Gaussian noise were received. The α-trimmed mean filter proved to be the most effective for both noise types.

Author details

Eugene Fedorov[1*], Tetyana Utkina[1] and Tetyana Neskorodeva[2]

1 Cherkasy State Technological University, Cherkasy, Ukraine

2 Vasyl' Stus Donetsk National University, Vinnytsia, Ukraine

*Address all correspondence to: fedorovee75@ukr.net

References

[1] Diniz PSR. Adaptive Filtering: Algorithms and Practical Implementation. Berlin: Springer; 2020. 505 p. DOI: 10.1007/978-1-4614-4106-9

[2] Lim JS. Two-Dimensional Signal and Image Processing. Englewood Cliffs, NJ: Prentice Hall; 1990. p. 694

[3] Rabiner LR, Schafer RW. Theory and Applications of Digital Speech Processing. Upper Saddle River, NJ: Pearson Higher Education, Inc.; 2011. p. 1042

[4] Yektaeian M, Amirfattahi R. Comparison of spectral subtraction methods used in noise suppression algorithms. In: Proceedings of 6th International Conference on Information, Communications and Signal Processing (ICICS 2007). 2007. pp. 1-4

[5] Mallat S. A Wavelet Tour of Signal Processing: Sparse Way. 3rd ed. Bourlington, MA: Academic Press; 2008. p. 832

[6] Gonzalez R, Woods R. Digital Image Processing. Hoboken, NJ: Pearson Education, Inc.; 2018. p. 1306